European Consortium for
Mathematics in Industry 2

Engl/Wacker/Zulehner (Eds.)
Case Studies in Industrial Mathematics

European Consortium for Mathematics in Industry

Edited by
Michiel Hazewinkel, Amsterdam
Helmuth Neunzert, Kaiserslautern
Alan Tayler, Oxford
Hansjörg Wacker, Linz

ECMI Vol. 2

Within Europe a number of academic groups have accepted their responsibility towards European industry and have proposed to found a European Consortium for Mathematics in Industry (ECMI) as an expression of this responsibility.

One of the activities of ECMI is the publication of books, which reflect its general philosophy; the texts of the series will help in promoting the use of mathematics in industry and in educating mathematicians for industry. They will consider different fields of applications, present casestudies, introduce new mathematical concepts in their relation to practical applications. They shall also represent the variety of the European mathematical traditions, for example practical asymptotics and differential equations in Britain, sophisticated numerical analysis from France, powerful computation in Germany, novel discrete mathematics in Holland, elegant real analysis from Italy. They will demonstrate that all these branches of mathematics are applicable to real problems, and industry and universities in any country can clearly benefit from the skills of the complete range of European applied mathematics.

Case Studies in
Industrial Mathematics

Edited by
Heinz W. Engl, Hansjörg Wacker, Walter Zulehner

With Contributions of
Dietmar Auzinger
Wolfgang Bauer
Heinz W. Engl
Thomas Langthaler
Ewald H. Lindner
Leopold Peer
Hansjörg Wacker
Walter Zulehner

 Springer Fachmedien Wiesbaden GmbH

Contributors:

o. Univ. Prof. Dr. Hansjörg Wacker
a. Univ. Prof. Dipl.-Ing., Dr. Heinz W. Engl
Dipl.-Ing. Dr. Walter Zulehner
Dipl.-Ing. Dietmar Auzinger
Dipl.-Ing. Dr. Wolfgang Bauer
Dipl.-Ing. Thomas Langthaler
Dipl.-Ing. Dr. Ewald H. Lindner
Dipl.-Ing. Leopold Peer
Institut für Mathematik
Johannes-Kepler-Universität Linz/Österreich

CIP-Titelaufnahme der Deutschen Bibliothek

Case studies in industrial mathematics / ed. by Heinz W. Engl...
With contributions of Dietmar Auzinger ...

   (European Consortium for Mathematics in Industry; Vol. 2)
     ISBN 978-3-663-12064-3   ISBN 978-3-663-12063-6 (eBook)
     DOI 10.1007/978-3-663-12063-6
NE: Engl, Heinz W. (Hrsg.); Auzinger, Dietmar (Mitverf.); European Consortium for Mathematics in Industry; European Consortium for ...

Library of Congress Cataloguing in Publication Data

CIP data appear on separate card.

Copyright © 1988 by Springer Fachmedien Wiesbaden
Originally published by B.G. Teubner, Stuttgart in 1988

Introduction

1 Industrial Mathematics in Linz

The Johannes-Kepler-University is situated in Linz, which is the industrial center of Austria. This location provides unique opportunities for cooperation between industry and a university which derives its name from one of the most eminent applied mathematicians of all times. The mathematics department was founded in the late Sixties, the first students graduated in 1974. In these boom times, they had no problems of finding jobs in industry. However, their employers were then more interested in their general training than in their specific mathematical skills. To change this, the department decided to actively seek cooperation with industry in what we called "problem seminars", where students were trained to solve (under guidance) real-world problems from local industry. Groundwork was already laid by a curriculum with special emphasis on numerical analysis, statistics, and optimization. This work in problem seminars usually evolved into diploma theses. Incidentally, in all projects presented here, students were involved at some stages. While the original motivation for cooperation with industry was educational, it turned out that most problems presented to us also led to interesting mathematical problems, so that nowadays our motivation is as much scientific as educational.

When cooperating with industry, one cannot expect that the problems to be solved fall into one's special mathematical interest. However, in order to guarantee the necessary scientific depth, we tried to concentrate on problems close to the mathematical expertise of our groups, which includes nonlinear equations, optimization and control, inverse and ill-posed problems. For six out of the eight projects presented here, the essential mathematical tools came from these fields. Another important point is that one has to build up a certain understanding of the field of applications one is dealing with. Because of this, we tried (after the initial phase where our motivation was mainly educational, so that a wide variety of not too hard problems from different areas was welcome) to concentrate on a few areas of applications, namely hydro energy production, various phases of steel production, modeling of large-scale chemical systems. This list is of course the consequence of the willingness of industrial partners in these fields to cooperate with us not just on one project, but on a more regular basis. Recently, computational fluid dynamics was added to this list, but this is not yet reflected in this volume.

As the use of the term "Case Studies" in the title indicates, all projects presented here are "real world" problems posed to us from industry and were treated in close contact with engineers. Since the final products the industrial partners were interested in were in all cases working computer programs (on a sound mathematical basis), numerical methods play a crucial role in all projects. Although standard software was also used in parts of some projects, it was essential for final success in each project to develop special solution techniques for some part of the problem. In some cases, this led to new mathematical results.

Each project also contained a modeling part, in which mathematicians from our groups were always involved. We found it instructive also to report about occasional failures in a first attempt of modeling or of a mathematical approach.

From our perspective, cooperation between mathematicians and engineers turned out to be fruitful in two ways. First, we profited from their deeper understanding of the practical side. But second, we think that we could contribute what might be considered typical "mathematical thinking". To illustrate this, we give some examples from our experience: When modeling phenomena, which a mathematician would model by a differential equation, engineers sometimes take a discrete model from the beginning. Besides giving less insight into qualitative aspects, this approach prescribes a specific solution technique and does not leave open the possibility of considering different discretizations or even other numerical methods for solving the differential equation. Also the role of optimization and of inverse problems is sometimes underestimated in the engineering community. Optimization is often replaced by simulation, the specific mathematical and numerical difficulties of inverse problems are not widely known. In these aspects, specific mathematical training is certainly useful and necessary.

2 The Projects to be Presented

As opposed to some other volumes on Industrial Mathematics (see Section 3), we present only projects treated by our own groups in Linz. This necessarily narrows the scope both in mathematical methods used and in application areas covered as indicated in Section 1. We now shortly describe the projects to be treated in this volume:

Project 1 In the design of reactor coils, which are large electrical components e.g. for limiting the current under system fault conditions, it is important to know the effect of eddy currents. The mathematical model leads to a combined boundary value and transmission problem for a combination of a Laplace and a Helmholtz equation. Because of the computer space limitations in our partner company it was necessary to develop a specific algorithm for this problem consisting of discretization, decomposition and boundary integral equations. An unsuccessful first attempt showed the importance of the question of well-posedness of the problem.

Project 2 The hydrodynamic coefficients of a body submerged in water quantitatively describe the influence of the surrounding water on the motion of the body. They are of practical value, e.g., for studying critical vibrations of submerged bodies, such as turbines. The calculation of the coefficients leads to a series of boundary value problems for Laplace's equation with von Neumann boundary conditions. For bodies of revolution, these problems were solved by a finite element method with rotationally symmetric elements.

Project 3 In continuous casting of steel, it is important both for quality and for technological reasons that one is able to control the boundary between the solid and the liquid phase. This can be done by approximately adjusting the water pressure in the secondary cooling zones. Mathematically, the question of how to set this water pressure is an inverse problem of Stefan type and hence very ill-posed. It turned out that for technological reasons, there was a natural "built-in regularization", which also allowed to reduce the problem to a finite-dimensional nonlinear optimization problem (where each evaluation of the objective functional involved the solution of a nonlinear boundary value problem).

Project 4 If the slabs of steel do not immediately enter the rolling process, after the casting, they cool and, therefore, have to be reheated before continuing the process. Reheating is assumed to be done by a pusher type furnace. Part 1 of the paper describes the optimization of the reheating process. A steady state model for the slabs, the flue gas and the recuperator is used. Mathematically, one obtains a coupled system of a nonlinear heat equation, some ordinary differential equations and a number of nonlinear equations. A sequential quadratic programming technique proposed by Schittkowski is used for the optimization. Part 2 contains a first attempt for an on-line control based on the results of Part 1. For a given optimal bulk temperature of the slabs throughout the furnace, the temperature in the furnace is controlled such that the bulk temperature (determined by a model) approximates the optimal bulk temperature. In order to study the influence of the skids on which the slabs are pushed through, a three-dimensional model is used.

Project 5 For designing the casing channel (volute) of a centrifugal pump either the principle of constant angular momentum (after C. Pfleiderer) or the principle of constant average speed (after A. J. Stepanoff) is often used. In order to improve the efficiency of the pump, both principles are combined in this work and, additionally, maximum hydraulic diameters are required for the cross sections of the volute. This leads to a classical problem in the calculus of variation with two constraints and a variable end point. The Euler-Lagrange equation and the transversality condition for this problem are reduced to a system of three nonlinear equations in three unknowns, which was numerically solved by Newton's method. Homotopy techniques were used to determine an initial guess.

Project 6 Distillation is one of the most common processes for separating a liquid or vapor mixture. In industry, it is usually carried out by a distillation column. A steady state model for such a column is described. This leads to a large scale system of nonlinear equations with sparse Jacobian. Solution techniques are developed under exploitation of the special structure of the problem. An industrial plant for carrying out more general separation processes is a more or less complex configuration of interlinked devices, such as distillation columns, absorbers and others. For the resulting nonlinear system a splitting technique is introduced, which reduces the complexity of the problem. Finally, a "real world" plant for producing oxygen from air is shortly studied.

Project 7 The optimization of a system of three hydro energy power plants is described. Some special features of the problem are that the storing capability of the plants is reduced by heavy seepage losses and that certain requirements related to tourism resp. a contract with the Austrian railway company have to be taken into account. Mathematically, one obtains a problem of optimal control with constraints both for the control and the state variables. A simplified model can be treated by variational techniques, the optimal solution is either bang-bang or singular. For the full problem, dynamic programming and a specially developed convexification technique proved to be successful for problems of about 1000 variables and 4000 constraints. For peak power demands, nonlinear programming was used. A considerable increase of production could be achieved.

Project 8 A special aspect which plays an important role in hydro energy optimization is the determination of the efficiency of a plant. After sketching two unsuccessful attempts, a new black-box technique is developed, which is based on a generalized nonlinear least squares method proposed by Schwetlick and Tiller. The essential difference compared with the usual least squares technique is that the points where the measurements are taken may also involve errors. Mathematically, one has to solve medium sized nonlinear problems (about 100 variables). By a suitable technique the linearized problem can be split into smaller problems depending on the number of measurements. This technique was successfully applied to determine the efficiency of two power plants in Upper Austria.

3 Bibliographical Notes on Industrial Mathematics

When editing this volume, we had three main purposes in mind: First, we want to convince engineers about the usefulness of cooperating with mathematicians when solving their problems. Second, we want to convince mathematicians that industrial problems offer (besides the satisfaction of cooperating with practise) challenging mathematical questions. Third, we want this book to be used e.g. in problem seminars with students. Before solving actual new problems, it might be helpful in a preparatory phase to rely on problems already treated by other groups and redo part of the modeling or numerical work. Of course, there are other books which can be used for all these purposes. We mention a few:

[6] is an excellent starting point for getting students interested in mathematical modeling, since some non-standard examples from engineering and natural sciences are given where mathematics is used in an interesting way. [2] discusses mathematical modeling techniques thoroughly in a general framework and then presents three case studies. Also [8] is a book about modeling in general, although more aimed at problems from physics and mechanics; five case studies are given.

[7] reflects the long-term experience of A. Taylor and his group in Industrial Mathematics, both educationally (the Oxford M.Sc.-program) and in actual industrial projects ("Oxford Study Groups"). The volume is organized along mathematical techniques, each one being illustrated by a practical problem. Mathematically, emphasis is laid on analytical methods, especially for partial differential equations.

More emphasis on numerical methods is laid in [1], [3], [4]. These books, which are organized not according to methods, but according to the practical problems treated, reflect the experience of Britain's "University Consortium in Industrial and Numerical Mathematics (UCINA)" and Australia's "Commonwealth Scientific and Industrial Research Organization (CSIRO)".

In [5], finally, one can find general articles about cooperation between universities and industry in mathematics, reports about individual projects and problems posed by industry. A forthcoming volume in this series, also edited by H. Neunzert as proceedings volume of an Oberwolfach conference, will include a large number of case studies in Industrial Mathematics from all over the world.

4 References

[1] R. S. Anderssen, F. R. de Hoog (eds.), The Application of Mathematics in Industry, Martinus Nijhoff, The Hague 1982
[2] R. Avis, Mathematical Modelling Techniques, Pitmann, London 1978

[3] N. G. Barton, J. D. Gray (eds.), Proc. of the 1985 Mathematics-in-Industry Study Group, CSIRO, Australia 1986

[4] C. M. Elliott, S. McKee (eds.), Industrial Numerical Analysis, Clarendon Press, Oxford 1986

[5] H. Neunzert (ed.), Mathematics in Industry, Teubner, Stuttgart 1984

[6] R. Seydel, R. Bulirsch, Vom Regenbogen zum Farbfernsehen, Springer, Berlin 1986

[7] A. B. Tayler, Mathematical Models in Applied Mathematics, Clarendon Press, Oxford 1986

[8] P. E. Wellstead, Physical System Modelling, Academic Press, London 1979

Acknowledgement

We gratefully acknowledge financial support for the projects described in this volume, which has been provided by the Austrian Fonds zur Förderung der wissenschaftlichen Forschung, the Jubiläumsfonds of the Austrian National Bank, the Austrian Federal Ministry for Science and Research (BMWF), the Oberösterreichische Kraftwerks AG (OKA), and the VOEST-ALPINE AG. Specifics and personal acknowledgements can be found at the end of the respective sections.

Table of Contents

COMPUTING EDDY CURRENT LOSSES IN REACTOR COILS

Heinz W.Engl and Ewald Lindner

1 The Problem

Reactor coils are electrical components which serve (among others) the following purposes: to limit the current under system fault conditions, to distribute the load to parallel circuits, to reduce the starting current that occurs when electrical machines are switched on, to limit short-circuit currents in rectifier equipments. In the design of reactor coils it is important to know how big the losses in current caused by eddy currents are; eddy currents are alternating currents induced in the conductors by the alternating magnetic field of the coil.

The problem posed to us by a Linz-based company that manufactures large reactor coils was to write a computer program that calculates these eddy current losses; the computer program should run on their (then) relatively small computer.

It turned out that before a working program could be written, a substantial amount of non-trivial mathematics had to be done. Actually, a first attempt to write a program before thinking about the mathematics of the problem failed.

The presentation in this paper is organized in such a way that it reflects the chronological order in which the different aspects of the problem were treated.

Details can be found in [6] and [9].

Of course, the underlying equations are Maxwell's equations together with material laws:

$$(1.1) \quad \text{curl } H = J + \frac{\partial D}{\partial t}$$

$$(1.2) \quad \text{curl } E = - \frac{\partial B}{\partial t}$$

(1.3) $\operatorname{div} (J + \frac{\partial D}{\partial t}) = 0$

(1.4) $\operatorname{div} B = 0$

(1.5) $D = \varepsilon . E$

(1.6) $\dot{J} = \sigma . E$

(1.7) $B = \mu . H .$

Here, the occuring vector fields have the following meanings:

H	magnetic field
E	electric field
D	dielectric displacement
B	magnetic induction
J	current density,

the constants are:

ε	absolute dielectric constant of the material
σ	specific conductivity
μ	absolute magnetic permeability of the material.

These equations have to be applied to any single conductor wound around the supporting cylinder of the reactor coil and to the surrounding air. From the outset, it is assumed that all physical constants are really constant in the conductor and in the air, respectively.

Using the vector potential A defined by

(1.8) $B = \operatorname{curl} A ,$

we obtain from (1.2) that

(1.9) $E = - \frac{\partial A}{\partial t} - \operatorname{grad} V ,$

where V is a scalar field. Assuming that $\operatorname{div} A = 0$, we obtain from (1.1), (1.7) and (1.8) that

(1.10) $\Delta A = - \mu (J + \frac{\partial D}{\partial t}) .$

With (1.6) and (1.9), this implies

(1.11) $\Delta A = \mu\sigma\frac{\partial A}{\partial t} + \mu\sigma \operatorname{grad} V - \mu\frac{\partial D}{\partial t} .$

Note that because of (1.8), (1.4) is fulfilled. Also, (1.1) implies (1.3). Thus, (1.11) (together with (1.8) and div $A = 0$) suffices (as soon as V and D are known) to describe all quantities of interest; the electric field can be computed from (1.9).

In the practical problem we were dealing with, we could make the following further simplifying assumptions:

Current flows only in z-direction, which is the direction of the conductor; all time-dependent quantities depend on time in a sinusoidal way with the same fixed frequency $\frac{\omega}{2\pi}$, which is between 50 and 500 Hz. J and B are independent of z. Thus, A has the form

(1.12) $A(x,y,z,t) = (0,0, \text{Re}(A(x,y) \cdot e^{i\omega t}))$.

The cross-section of the conductor is a rectangle with width b between 2 and 5 mm and lenght h between 1 and 2 cm. Thus, the dimensions of the conductor are much smaller than the wave-lenght of the electric field, so that the density of the displacement current is negligible (cf.[14,p.368]). Thus, we set $\frac{\partial D}{\partial t} = 0$ in all equations above.

grad V depends (by the assumptions) on (x,y) only. On the other hand, V depends on z only, since otherwise, by (1.9), an electric current in the cross-section of the conductor would be generated contrary to our assumption. Together, this implies that grad V has the form

(1.13) grad $V = (0,0,-\frac{1}{\sigma}J_s)$,

where J_s is constant in the conductor and in the air, respectively (cf.[15]). The "source current density" J_s has values between 10^6 and $2*10^6$ Am^{-2} in the conductor and vanishes in the air. With the special form of A assumed in (1.12) and these considerations leading to (1.13), (1.11) reduces to the two-dimensional scalar equation

(1.14) $\Delta A = i\omega\sigma\mu A-\mu J_s$.

The quantity that is of interest, namely the eddy current loss P, follows from A by

(1.15) $P = \frac{1}{2\sigma} \int_I |i\omega\sigma A(x,y) - J_s|^2 \, d(x,y)$

4

(cf. [15]), where I is the cross-section of the conductor.
Note that (1.14) holds in the conductor and in the surrounding
air (with different values for σ, μ and J_s). Although A is needed in (1.15) in the conductor only, (1.14) has to be considered
also in the surrounding air, since no boundary conditions on the
boundary of the conductor are known. This leads to the main difficulty: (1.14) is a Helmholtz equation in the conductor, but a
Laplace equation in the air; both a coefficient and the inhomo-
geneity of the equation have jump discontinuities at the bound-
dary of the conductor. If we denote by

$$(1.16) \quad \kappa^2 := -i\omega\sigma\mu,$$

then we obtain the following transmission problem (with $u := A$,
$f := -\mu J_s$):

$$(1.17) \quad \Delta u + \kappa^2 u = f \qquad \text{in } I$$
$$(1.18) \quad \Delta u = O \qquad \text{in } E$$
$$(1.19) \quad u^+ = u^- \qquad \text{on } IB$$
$$(1.20) \quad \frac{\partial u^+}{\partial n} = \frac{\partial u^-}{\partial n} \qquad \text{on } IB$$
$$(1.21) \quad \frac{\partial u^-}{\partial n} = g \qquad \text{on } EB.$$

Here, $E \subseteq \mathbb{R}^2$ denotes a bounded domain surrounding the
cross-section I of the conductor with boundary EB; IB is the
boundary of the conductor. More precisely: $G, I \subseteq \mathbb{R}^2$ are open,
bounded and simply connected regions, $\bar{I} \subseteq G$, $E := G \backslash \bar{I}$. $\frac{\partial}{\partial n}$ denotes
the (outward) normal derivative, the superscripts + and − stand
for the limits obtained from approaching IB and EB from the ex-
terior or interior, respectively. The transmission condition
(1.20) follows from the assumption (which is a good approximation of reality, cf.[8,p.219]) that μ is the same in the air and
in the conductor. In practice, E is chosen in such a way that
$\frac{\partial u^-}{\partial n}$ (which, by (1.8) and (1.12), is a component of the magnetic
field) can be measured on its boundary, which leads to (1.21).
Usually, the linear dimensions of E are about twice (and up to
four times) those of I.

This point of view is slightly different from that in
[4]: We assume that on EB we can measure a certain component of

the total field, while in [4], the knowledge of an incident
field (which is scattered by the conductor) in the interface
between conductor and air is assumed.

As already mentioned, the aim was to write a computer
program to compute P as given by (1.15), where A:=u is determined by (1.17) - (1.21). In order to do this, one has to under-
stand the mathematics of the problem. Especially, it is important to know if (1.17) - (1.21) is well-posed, since if one
deals with an ill-posed problem, a numerical approach not taking into account the ill-posedness will almost certainly fail;
cf. the article by Engl and Langthaler in this volume for aspects of ill-posed problems. In the next section, we deal with
the question of well-posedness.

2   Mathematics of the Problem

According to Hadamard, problem (1.17) - (1.21) is well-
posed, if for all data f and g, a unique solution u exists that
depends continuously on the data. In order to give this a mathematical meaning, we have to fix function spaces for the data
functions f and g and for the solution u. Since we look for
classical solutions, it turns out to be appropriate (see below)
to assume that f and g are (complex-valued) Hölder continuous
functions on $\overline{I}$ and EB, respectively, and to look for solutions
u that are continuously differentiable on $\overline{I}$ and $\overline{E}$ and twice continuously differentiable on I ∪ E. In this context, the problem
will turn out to be well-posed, where "continuous dependence" is
to be understood with respect to a Hölder norm on the data space
and the uniform norm on the solution space.

Of course, it would be desirable to base the proof of
well-posedness on a method that could also be used as a basis
for a numerical procedure. We achieved this at least to some
extent in [6]; in the meantime, the gap between theory and practice still present in [6] has been closed in [16]. For the numerical solution of (1.17) - (1.21), the following alternatives
are available: finite differences, finite elements, and a boundary integral equation method. Since, as mentioned above, the
computer at the company which posed the problem had quite limi-

ted memory space, it was crucial to decide which type of approach to take under the aspect of storage space. From this aspect, a boundary integral equation method seems to be most attractive because of the resulting reduction in the dimension of the problem. A boundary integral equation method (and its numerical implementation in the form of the "boundary element method") relies on the explicit knowledge of a fundamental solution of the underlying equation, which limits its applicability. Since for (1.17) and (1.18), fundamental solutions are known (cf. [3]), this method is in principle applicable here. The problem is that we have to work with two different equations.

Finite element and finite difference methods have the disadvantage over the boundary integral equation method that the resulting systems of equations are larger, which already ruled out their exclusive use in our problem. As discussed in Section 3, we used a combination of a boundary integral equation method in E and a finite difference method in I together with a special decoupling of the problem (1.17) - (1.21) into two separate problems in E and I, respectively.

In [7], a related eddy current problem in an unbounded domain is treated by a boundary integral equation method; for unbounded domains, boundary integral equation methods offer the additional advantage that the unbounded domain is reduced to its boundary; also [1] uses a boundary integral equation method, while in [2], [5], [12], finite elements are used. In [13], both methods are compared. However, as opposed to our problem, all these papers except [7] work with a single domain without transmission conditions. In [10], a 3-dimensional eddy current problem in the form of a transmission problem is treated by an integral equation method, which is also used for an asymptotic analysis ("skin effect approximation", cf. also [11]).

In addition to the reduction in dimension, boundary integral equation methods also have the advantage that they lead to existence proofs, usually via the Riesz-Schauder theory for compact operators (cf. [3]) or via the theory of pseudodifferential operators, if the integral operators involved are no longer

compact as it happens in the case of polygonal domains (see
[4]). Boundary integral equations for PDE's can be derived in at least two different ways: from the "3rd Green's
identity" (see below) and from the jump relations of single and
double layer potentials (cf.[3]). If one aims at integral equations of the second kind (because of the applicability of the
Riesz-Schauder theory), one should use single (double) layer
potentials for Neumann (Dirichlet) problems. Thus, looking at
(1.17) - (1.21), the first idea is to derive an equivalent
system of integral equations by using single layer potentials
with the fundamental solution of Laplace's equation on IB and
on EB and a double layer potential with the fundamental solution
of the Helmholtz equation (1.17) (which is a Hankel function).
This approach leads to a system of three integral equations,
whose unknowns are the densities of the three potentials on IB
and EB. Hence, the desired reduction in dimension can be a-
chieved in this way. These integral equations have been derived
in [9,p.77 ff.]. However, since in the process of using this
approach to prove existence of a solution we arrived at a non-
compact (even unbounded) operator (namely, the normal derivative
of the double layer potential for the Helmholtz equation, i.e.,
the two-dimensional analogue of the operator defined in (2.81) of
[3]), we abandoned this approach.

The next idea, which was eventually carried through in [9],
[6] (and will be outlined here) was to use potentials generated
by the fundamental solution of Laplace's equation only, since
the double layer potential for the Helmholtz equation caused the
difficulties in the first approach. As we will see, the price to
pay is that we have to use also a volume potential, so that the
desired reduction of dimension cannot be fully achieved. Fortunately, this volume potential is defined on the smaller domain
I.

Let

$$(2.1) \quad \gamma(x,y) := -\frac{1}{2\pi} \log |x-y|$$

for $x \neq y \in \mathbb{R}^2$ be the fundamental solution of Laplace's equation
in $\mathbb{R}^2$. For any bounded measurable subset B of $\mathbb{R}^2$ with piecewise C^1-boundary ∂B, we define the following potentials:

the "volume potential" with continuous density $\psi:\overline{B} \to \mathbb{C}$ by

(2.2) $\quad V(\psi,B)(x) := \int_B \psi(y)\gamma(x,y)dy,$

the "single layer potential" with continuous density

$\psi:\partial B \to \mathbb{C}$ by

(2.3) $\quad S(\psi,\partial B)(x) := \int_{\partial B} \psi(y)\gamma(x,y)\ ds(y),$

and the "double layer potential" with continuous density

$\psi:\partial B \to \mathbb{C}$ by

(2.4) $\quad D(\psi,\partial B)(x) := \int_{\partial B} \psi(y)\frac{\partial}{\partial n(y)}\gamma(x,y)\ ds(y),$

where $\frac{\partial}{\partial n(y)}$ denotes the normal derivative with respect to the variable y. The relevant properties of these potentials are summarized in [6].

The "3rd Green's identity" mentioned above, which will be used for deriving the integral equation, has the form

$$\eta(x)u(x) = -\int_B (\Delta u)(y)\gamma(x,y)\ dy - \int_{\partial B} u(y)\frac{\partial}{\partial n(y)}\gamma(x,y)\ ds(y) +$$

(2.5) $$+ \int_{\partial B} \frac{\partial u}{\partial n}(y)\gamma(x,y)\ ds(y),$$

with $u \in C^1(\overline{B}) \cap C^2(B)$ such that $\Delta u \in L^\infty_o(B)$, $x \in \mathbb{R}^2 \setminus \partial B$. The factor $\eta(x)$ is 0 for $x \in \mathbb{R}^2 \setminus \overline{B}$ and 1 for $x \in B$. Using the potentials defined above, (2.5) reads

(2.6) $\quad \eta u = -V(\Delta u,B) - D(u,\partial B) + S(\frac{\partial u}{\partial n},\partial B).$

Now assume that u solves (1.17) - (1.21) and let

(2.7) $\quad v_1 := u|_{\overline{I}}\ ,\quad v_2 := u|_{EB},\quad v := (v_1,v_2).$

Now we apply (2.6) with $B:=I$ for $x \in I$ and obtain with (1.17)

(2.8) $\quad u(x) = -V(f - \kappa^2 u,I)(x) - D(u,IB)(x) + S(\frac{\partial u}{\partial n},IB)(x);$

applying (2.6) with $B:=E$ and $x \in I$ yields with (1.18) that

(2.9) $\quad 0 = -D(u,IB \cup EB)(x) + S(\frac{\partial u}{\partial n},IB \cup EB)(x).$

Note that, since all normals are outward normals, the normals on IB appearing in the double layer potentials and in $\frac{\partial u}{\partial n}$ point into E in (2.8) and into I in (2.9). Taking this into account, we obtain by adding (2.8) and (2.9) together with (1.19) - (1.21) that

(2.10) $\quad u(x) = -V(f - \kappa^2 u,I)(x) - D(u,EB)(x) + S(g,EB)(x)$

holds for $x \in I$. Since all these potentials are continuous on $\overline{I}$

(note that in the last two potentials, the integration takes place over EB, which is not the boundary of I), (2.10) holds also for $x \in IB$. By (2.7), (2.10) implies that

(2.11) $v_1 - \kappa^2 V(v_1, I) + D(v_2, EB) = -V(f, I) + S(g, EB)$

holds (on $\bar{I}$).

Again by the continuity of the potentials at least on $\mathbb{R}^2 \setminus EB$, (2.10) holds also for $x \in E$. Now, V and S are continuous everywhere, while for D, we have the jump relation

(2.12) $\lim_{t \to 0^+} D(\psi, \partial B)(x \pm tn(x)) = D(\psi, \partial B)(x) \pm \frac{1}{2}\psi(x)$ $(x \in \partial B)$.

Thus, if we let x tend to EB from E (in normal direction), then all expressions in (2.10) except D(u,EB) remain formally the same, while D(u,EB) has to be replaced by $D(u,EB) - \frac{1}{2}u$. Hence, we have

(2.13) $\frac{1}{2}u(x) = -V(f - \kappa^2 u, I)(x) - D(u, EB)(x) + S(g, EB)(x)$

for $x \in EB$, which implies by (2.7) that

(2.14) $v_2 - 2\kappa^2 V(v_1, I) + 2D(v_2, EB) = -2V(f, I) + 2S(g, EB)$

holds on EB. (2.11) and (2.14) are the desired integral equations; explicitly, this system reads

$$(2.15) \quad \begin{aligned} &v_1(x) - \kappa^2 \int_I v_1(y)\gamma(x,y)dy + \int_{EB} v_2(y)\frac{\partial}{\partial n(y)}\gamma(x,y)ds(y) = \\ &= -\int_I f(y)\gamma(x,y)dy + \int_{EB} g(y)\gamma(x,y)ds(y) \quad (x \in \bar{I}), \end{aligned}$$

$$(2.16) \quad \begin{aligned} &v_2(x) - 2\kappa^2 \int_I v_1(y)\gamma(x,y)dy + 2\int_{EB} v_2(y)\frac{\partial}{\partial n(y)}\gamma(x,y)ds(y) = \\ &= -2\int_I f(y)\gamma(x,y)dy + 2\int_{EB} g(y)\gamma(x,y)ds(y) \quad (x \in EB), \end{aligned}$$

where γ is defined by (2.1).

So far we have shown that any solution of (1.17) - (1.21) generates a solution of (2.15) - (2.16) via (2.7). Conversely, it is shown in [6] that if (v_1, v_2) solves (2.15) - (2.16) then u defined by

$$(2.17) \quad u(x) := \begin{cases} v_1(x) & x \in \overline{I} \\ V(\kappa^2 v_1 - f, I)(x) + S(g, EB)(x) - D(v_2, EB)(x) & x \in E \\ v_2(x) & x \in EB \end{cases}$$

solves (1.17) - (1.21). In this sense, the two problems are equivalent. Furthermore , the process of defining u by (2.17) is Lipschitz-continuous, if the norms used for measuring errors in f, g, v_1, v_2 are Hölder norms and the uniform norm is used for u.

Thus, for proving well-posedness of (1.17) - (1.21), it suffices to show that (2.15) - (2.16) is well-posed in a space of Hölder continuous functions. This is achieved in [6] by re-writing (2.15) - (2.16) as an operator equation

$$(2.18) \quad (I-K)(v_1, v_2) = R(f, g)$$

on the space $C^{0,\alpha}(\overline{I}) \times C^{0,\alpha}(EB)$ with $0 < \alpha < 1$, where R symbolizes the right-hand sides of (2.11) and (2.14). It is then shown that K is compact and that $N(I-K) = \{O\}$. Thus, it follows from the Riesz-Schauder theory, that (I-K) is continuously inver-tible, so that (2.18) and hence (2.15) - (2.16) is well-posed in $C^{0,\alpha}(\overline{I}) \times C^{0,\alpha}(EB)$. All this works if $\int_{EB} \psi(y) \gamma(x,y) ds(y) = O$ has no nontrivial continuous solution, which is a (mild) restriction on E, and if IB is piecewise C^1 and EB is C^2.

Thus, under these assumptions we have:

<u>Theorem:</u> For each (f,g) $C^{0,\alpha}(\overline{I}) \times C^{0,\alpha}(EB)$, (1.17) - (1.21) has a unique solution u. The map that assigns each (f,g) this unique solution u is continuous from $C^{0,\alpha}(\overline{I}) \times C^{0,\alpha}(EB)$ to $C^0(\overline{IUE})$.

From this Theorem, we know that we are dealing with a well-posed problem. Only now it is safe to try to design a nu-merical algorithm, which we describe in the next section.

3   <u>Numerical Treatment of the Problem</u>

Due to the limitation in storage space and to the pro-blems mentioned above arising from the fact that the equations in E and I are different, we aimed at some sort of decoupling of the problem into separate problems for Laplace's and Helmholtz' equation; the idea was that these separate problems would contain some free parameter (function) to be determined by the transmission condition.

The first attempt was a failure; we think it is quite instructive to report about this failure since it is an example of what can happen if one does not think about the mathematics of the problem before doing numerical work.

This first idea was to decouple (1.17) - (1.21) into the following two problems:

u_1 should be a solution of

$$(3.1) \quad \Delta u_1 = 0 \qquad \text{in } E \cup \bar{I}$$

$$(3.2) \quad \frac{\partial u_1^-}{\partial n} = g \qquad \text{on } EB,$$

while u_2 should solve

$$(3.3) \quad \Delta u_2 + \kappa^2 u_2 = f - \kappa^2 u_1 \qquad \text{in } I$$

$$(3.4) \quad u_2 = 0 \qquad \text{in } \bar{E}.$$

Thus, $u_1 + u_2$ would solve (1.17) - (1.19), (1.21). To satisfy (1.20), we wanted to use the non-uniqueness of the solution of (3.1), (3.2); in fact, that problem contains one free parameter. Since (1.17) - (1.21) is solvable, it seemed that (1.20) could be satisfied by determining this one parameter. Of course, this was a bit suspicious, since how could one satisfy the infinitely many conditions contained in (1.20) by choosing just one free parameter correctly? Nevertheless , we did computations based on this approach using programs from the NAG-library. The results were a disaster!

The reason for this was of course that while no condition has to be imposed on g to guarantee the solvability of (1.17) - (1.21), (3.1) - (3.2) is solvable only if $\int_{EB} g \, ds = 0$.

Although this mistake we made is quite trivial, it still shows that one can avoid numerical effort leading nowhere by thinking about the well-posedness of the problem before doing calculations, as we have done in Section 2.

Nevertheless , the basic idea of decoupling proved to be feasible in the following way:

Let $h: IB \to \mathbb{C}$ be a (suitably smooth) function (to be determined later) such that

$$(3.5) \quad \int_{IB} h\,ds = \int_{EB} g\,ds$$

holds. The problem (1.17) - (1.21) is now decoupled into the following subproblems:

<u>Subproblem 3.1:</u> Let $u_I(h) \in C^1(\overline{I}) \cap C^2(I)$ be the solution of (1.17) with

$$(3.6) \quad \frac{\partial u^-}{\partial n} = h \qquad \text{on } IB.$$

<u>Subproblem 3.2:</u> Let $x_0 \in IB$, $u_I(h)$ be the solution of Subproblem 3.1. Let $u_E(h) \in C^1(\overline{E}) \cap C^2(E)$ be the solution of (1.18), (1.21),

$$(3.7) \quad \frac{\partial u^+}{\partial n} = h \qquad \text{on } IB$$

$$(3.8) \quad u(x_0) = u_I(h)(x_0).$$

Now, let

$$(3.9) \quad u(h): \overline{EUI} \;\to\; \mathbb{C}$$
$$x \;\to\; \begin{cases} u_I(h)(x) & x \in I \\ u_E(h)(x) & x \in \overline{E}. \end{cases}$$

Then, $u(h)$ fulfills (1.17) - (1.18), (1.20) - (1.21). The parameter function h now has to be determined in such a way that also (1.19) is fulfilled. This leads to

<u>Subproblem 3.3:</u> Find h fulfilling (3.5) such that with $u(h)$ de-defined by (3.9), (1.19) holds.

This approach has subsequently been studied in [16] and has there been carried over to more general transmission conditions than (1.19), (1.20) and to the situation that in E and I, we have Helmholtz equations with different values for κ. A slight (but insignificant) difference to our situation is that in [16], E is unbounded and hence, (1.21) is replaced by the Sommerfeld radiation condition. In [16], it is outlined how this decoupling approach can be used not only for numerical purposes, but for proving well-posedness. This bridges the gap between theory and practice still present in [6]. We remark that because of (1.16), Im κ > O, so that Subproblem 3.1 has a unique solution. Because of (3.5) and (3.8), Subproblem 3.2 also has a unique solution. Because of the results of Section 2, also Subproblem 3.3 is uniquely solvable.

The next question is how to solve these subproblems numerically . Of course, as far as storage space is concerned, it would be optimal to solve Subproblem 3.1 by a boundary integral equation method. We did not do so, since we had no library program for computing the necessary Hankel functions, which appear as fundamental solutions of (1.17) in the kernel of the integral. However, we did solve Subproblem 3.2 by a boundary integral equation method based on (2.5), since there, the fundamental solution is given by (2.1). The resulting integral equation was solved by a Galerkin collocation method with piecewise constant functions as basis functions. The reason why we chose piecewise constant functions was that in this way, the coefficients of the resulting linear system, which involve integrals of the type $\int_{P_i} \frac{\partial}{\partial n(y)} \log |x-y| ds(y)$ (where the P_i are pieces of a

polygon approximating EBUIB), could be calculated explicitly. Without a computer algebra system, the explicit evaluation of the corresponding integrals with, say, linear splines as basis functions seemed to be much too involved. The use of numerical quadrature would be less accurate.

The linear system resulting from (1.18), (1.21), (3.7) is underdetermined; we computed its solution of minimal norm with Householder's method. Then, a suitable constant was added to fulfill (3.8). Note that since $u_I(h)$ appears in (3.8), Subproblem 3.1 had to be solved first. We did this by a finite-difference method. Because of the limitations in storage space, we had to use the sparsity of the resulting linear system.

Because of the dimensions of I (i. e., the cross-section of the conductor), we had to discretize with different discretization parameters in x-and y-directions; the Laplacian was approximated by a five-point rule on this non-uniform grid.

The matrix of the resulting linear system was block-tri-diagonal, each of the blocks (most of them identical) was a tri-diagonal matrix. The diagonal elements were complex numbers (because of (1.16)).

On this matrix, we performed an LU-decomposition by a block Gauß method together with successive approximation and iterative refinement. For the (quite lengthy) convergence and stability analysis see [9]. Independent tests for this algorithm for Subproblem 3.1 showed that the combination of the block Gauß method with successive approximation, which allowed to perform this block Gauß method only until two consecutive subdiagonal blocks were "sufficiently close", saved considerable computing time.

We now turn to Subproblem 3.3. Here, we approximated the unknown function h by linear splines h_0, h_1 ..., $h_n : IB \to \mathbb{C}$ in the form

$$(3.10) \quad h \approx h_0 + \sum_{i=1}^{n} \alpha_i h_i,$$

where the basis functions were chosen such that h_0 fulfilled (3.5),

$$(3.11) \quad \int_{IB} h_i \, ds = 0 \text{ for } i \in \{1, \ldots n\},$$

and $h_1, \ldots, h_n$ are linearly independent; let $x_1, \ldots, x_n$ be the knots for these splines. By our assumptions, the matrix

$$(3.12) \quad (h_i(x_j))_{i,j \in \{1, \ldots, n\}}$$

is regular.

Now, we determine the coefficients $\alpha_1 \ldots, \alpha_n$ in (3.10) such that (1.19) is fulfilled at the points $x_1, \ldots, x_n$, i. e.,

$$(3.13) \quad u_E (h_0 + \sum_{i=1}^{n} \alpha_i h_i)(x_j) = u_I (h_0 + \sum_{i=1}^{n} \alpha_i h_i)(x_j) \quad (j \in \{1, \ldots, n\})$$

holds, where u_I and u_E are the solutions of Subproblems 3.1 and 3.2, respectively.

If one characterizes the solutions of Subproblems 3.1 and 3.2 via boundary integral equations, then (1.19) and hence Subproblem 3.3 can be viewed as an integral equation for h; (3.13) can then be interpreted as a spline collocation method

for solving this integral equation, which converges to the
exact solution h of Subproblem 3.3 because of (3.12) and the
convergence of spline interpolants (cf. [16]).

(3.13) can be rewritten as a linear system of equations
for $\alpha_1, \ldots, \alpha_n$ as follows:

For $i \in \{1, \ldots, n\}$, let $\tilde{u}_I(h_i)$ and $\tilde{u}_E(h_i)$ be the solutions of
Subproblems 3.1 and 3.2 with h_i instead of h and f,g both repla-
ced by 0, respectively. Because of the linearity of Subproblems
3.1 and 3.2, (3.13) is equivalent to

$$u_E(h_0)(x_j) + \sum_{i=1}^{n} \alpha_i \tilde{u}_E(h_i)(x_j) =$$

(3.14)

$$= u_I(h_0)(x_j) + \sum_{i=1}^{n} \alpha_i \tilde{u}_I(h_i)(x_j) \quad (j \in \{1, \ldots, n\}),$$

i. e.,

(3.15) $A\alpha = b$,

where the coefficients a_{ij} of A are determined by

(3.16) $a_{ij} = \tilde{u}_E(h_i)(x_j) - \tilde{u}_I(h_i)(x_j) \quad (i,j \in \{1, \ldots, n\})$

and b is determined by

(3.17) $b_j = u_I(h_0)(x_j) - u_E(h_0)(x_j) \quad (j \in \{1, \ldots, n\})$.

Thus, for computing A and b, one has to solve Subproblems 3.1
and 3.2 repeatedly with different boundary values h_i, f,g .
Note that this leads to linear systems which differ only in
the right-hand sides, so that the amount of numerical work is
tolerable.

Once A and b are computed, one solves (3.15) (which is
uniquely solvable, cf. [16]) for α , defines h by (3.10) and
solves Subproblems 3.1 and 3.2 with this h. Note that for this
last step, the resulting linear systems differ from those used
in computing A and b only in the right-hand sides. Alternative-
ly, if one chose to store $u_E(h_0)$, $u_I(h_0)$, $\tilde{u}_E(h_i)$, $\tilde{u}_I(h_i)$, which
were already used, at all grid points (which we did not do), then
one could simply compute $u_E(h)$ as $u_E(h_0) + \sum_{i=1}^{n} \alpha_i \tilde{u}_E(h_i)$ and $u_I(h)$
analogously.

Now, u(h) defined by (3.9) serves as approximate solution for (1.17) - (1.21).

A count of multiplications gives the following result, where n denotes the maximum number of discretization points in x- or y- direction:
If we had discretized the system of integral equations (2.15) - (2.16) used in the proof of well-posedness, we would have ended up with a linear system of $O(n^2)$ unknowns (because of the presence of a volume potential!); this system would have been non-sparse, so that we would have needed $O(n^6)$ operations. In our approach, we still have $O(n^2)$ unknowns because of the use of a finite difference method for Subproblem 3.1; the use of the sparsity of the matrix as described above results in an operation count of $O(n^4)$. If we had used a boundary integral equation method also for Subproblem 3.1, we would have ended up with $O(n^3)$ operations (cf. [16]).

We close with two numerical examples; in both, I is a rectangular conductor $[0,2b] \times [0,2h]$, the linear dimensions of E, which is a parallel rectangle, are twice those of I in the first example, four times those of I in the second example. We assume the presence of a transverse magnetic field $B_O = 0.1$ V s m^2 in negative x-direction, which determines g. The constants have the following values:
$\omega = 100\pi$ s^{-1}, $\sigma = 36*10^6$ $AV^{-1}m^{-1}$, $\mu_O = 4\pi 10^{-7}$ $VsA^{-1}m^{-1}$, $J_s = 0$; κ is determined by (1.16).

<u>Example 1:</u> $b = 5*10^{-3}m$, $h = 2.5*10^{-3}m$. By a "rule of thumb" given in [15,p.120], P as defined by (1.15) should be approximately 1.849 VAm^{-1}. Our method, using in addition symmetry in y-direction and 4 discretization points in each direction in I, gave a value of $P = 1.8255$ VAm^{-1} and consumed 117 K byte storage space and 69 sec CPU-time on an IBM 360/155.

<u>Example 2:</u> $b = 1.19*10^{-2}m$, $h = 1.78*10^{-2}m$. Here, the "rule of thumb" mentioned above is not applicable any more. Using 4 and 6 discretization points in each direction in I, we obtained approximate values for P of 999 and 978 using 243 K and 477 K bytes storage space and 111 sec and 277 sec CPU-time, respectively. Linear extrapolation yields a value of $P = 935.7$ VAm^{-1}, which is

in good accordance with the value 932.30 given in [15,p.120] for slightly different values of the constants.

In [9], [6], more examples can be found. They all suggest linear convergence of the whole algorithm, which is proven in [16]. This also justifies the linear extrapolation in Example 2.

References:

[1] B. Ancelle, A. Nicolas, J. Sabonnadiére,
     A boundary integral equation method for high frequency
     eddy currents,
     IEEE MAG-17 (1981), 2568 - 2570

[2] M. Chiampi, A. Negro, M. Tartaglia,
     Finite elements computation of magnetic fields in multi-
     ply connected domains,
     IEEE MAG-17 (1981), 1493 - 1497

[3] D. Colton, R. Kreß,
     Integral Equation Methods in Scattering Theory,
     Wiley, New York 1983

[4] M. Costabel, E. Stephan,
     A direct boundary integral equation method for transmis-
     sion problems, Jour. Math. Analysis Appl. 106 (1985),
     367 - 413

[5] N. Demerdash, O. Mohammed, T. Nehl, F. Fonad, R. Miller,
     Solution of eddy current problems using three-dimensio-
     nal complex magnetic vector potential,
     IEEE PAS-101 (1982), 4222 - 4229

[6] H. W. Engl, E. Lindner,
     A combined boundary value and transmission problem ari-
     sing from the calculation of eddy currents: well-posed-
     ness and numerical treatment, J. of Appl. Math. and Phys.
     (ZAMP) 35 (1984), 289 - 307

[7] S. Hariharan,
     An integral equation procedure for eddy current problems,
     PhD-Thesis, Carnegie-Mellon University, Pittsburgh 1980

[8] K. Küpfmüller,
     Einführung in die theoretische Elektrotechnik,
     8. Auflage, Springer, Berlin 1965

[9] E. Lindner,
     Lösungstheorie und Numerik für eine partielle Differen-
     tialgleichung im Zusammenhang mit der Berechnung von

Wirbelströmen, Diplomarbeit, Johannes-Kepler-Universität, Linz 1983

[10] R. C. MacCamy, E. Stephan,
Solution procedures for three-dimensional eddy current problems, Jour. Math. Analysis Appl. 101 (1984), 348 - 379

[11] R. C. MacCamy, E. Stephan,
A skin effect approximation for eddy current problems, Arch. Rat. Mech. Analysis 90 (1985), 87 - 98

[12] K. Preis, H. Stögner, K. Richter,
Calculation of eddy current losses in air coils by finite element methods,
IEEE MAG-18 (1982), 1064 - 1066

[13] S. Salon, J. Schneider,
A comparison of boundary integral and finite element formulations of eddy current problems,
IEEE PAS-100 (1981), 1473 - 1479

[14] W. Smythe,
Static and Dynamic Electricity, 3rd ed., McGraw-Hill, New York 1968

[15] R. Stoll,
The Analysis of Eddy Currents,
Clarendon Press, Oxford 1974

[16] A. Zinn,
Eine numerische Methode zur Lösung eines Transmissionsproblems bei der Helmholtzgleichung, Diplomarbeit, Georg-August-Universität, Göttingen 1987

CALCULATION OF THE HYDRODYNAMIC COEFFICIENTS

FOR BODIES OF REVOLUTION

Walter Zulehner

1 Introduction

This project was initiated by the local branch of the steel company VOEST-ALPINE AG. The author of this paper wrote his diploma thesis on the subject, see [3].

Comparing the motion of a rigid body through air and through water, two significant differences are observed:

a. The body seems to be lighter in water, i.e. the body seems to have <u>less weight</u>. This is the well-known buoyancy effect.

b. The force necessary to accelerate a body in water is larger than in air, i.e. the body seems to have <u>more inertia</u>. E.g., for a simple one-dimensional translational motion, the mass of a body seems to have increased by some amount, usually called added mass or hydrodynamic mass. For a general motion, the effect of the surrounding water is expressed in terms of a virtually increased kinetic energy of the body, represented by a number of so-called hydrodynamic coefficients (including the hydrodynamic mass).

This paper deals with the second phenomenon. The calculation of the hydrodynamic coefficients is of practical value for studying vibrations of bodies submerged in water (e.g. turbines). The effect of the surrounding water on the vibration of a body is (almost) completely represented by these coefficients. Roughly speaking, the body submerged in water vibrates like a body in air with increased inertia.

In the following section the physical problem is described. The liquid is modeled by a potential flow. The hydrodynamic coefficients are expressed in terms of certain potentials. Section 3 contains a short analysis of the resulting mathematical problem, namely the Laplacian equation with von Neumann boundary conditions. For bodies of revolution, the problem is two-dimensional if cylindrical coordinates are introduced. In Section 4 the numerical method, a finite element technique, is presented and results of a convergence analysis are reported. The paper concludes with some numerical experiments.

2   The physical model

The motion of a solid through a liquid is a classic problem in fluid dynamics. This section mainly follows the presentation by Lamb [1].

2.1   The equations of motion for the rigid body

At each time t the rigid body is represented by a subset $G_2(t)$ of $\mathbb{R}^3$, the three-dimensional Euclidean space.

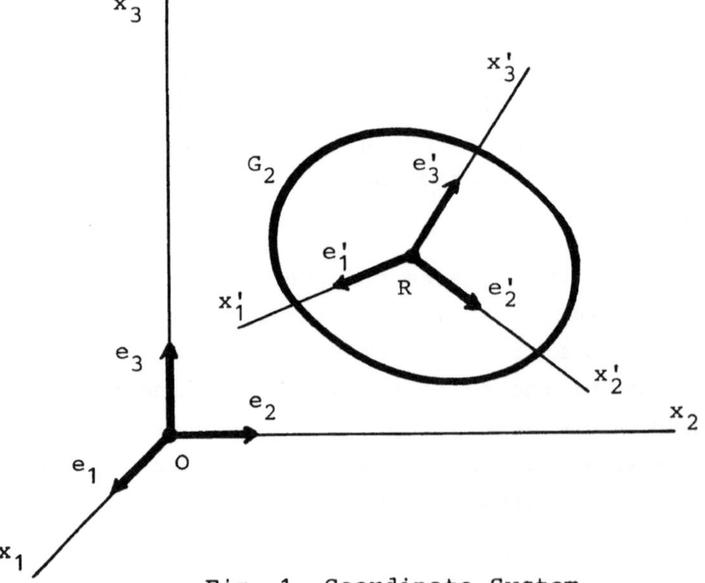

Fig. 1   Coordinate System

In addition to the coordinate system fixed in space (origin O, basic unit vectors e_1, e_2, e_3, coordinates x_1, x_2, x_3), a second Cartesian coordinate system fixed in the rigid body is introduced (origin R(t), basic unit vectors $e_1'(t)$, $e_2'(t)$, $e_3'(t)$, coordinates x_1', x_2', x_3').

The location of the rigid body is completely specified by the position of the origin R(t) and the matrix D(t) which transforms the system (e_1, e_2, e_3) to $(e_1'(t), e_2'(t), e_3'(t))$. The transformation matrix D(t) describes a rotation about some axis. The angular velocity $\omega(t)$ of the rigid body and the matrix D(t) are related by the equation

$$\frac{d}{dt}(D(t)x) = \omega(t) \times (D(t)x) \quad \text{for each } x \in \mathbb{R}^3$$

(the symbol $\times$ denotes the cross product in $\mathbb{R}^3$). The velocity of the rigid body at some point x and time t is given by

$$V(x,t) = U(t) + \omega(t) \times (x - R(t)),$$

where $U(t) = \frac{d}{dt}R(t)$ is the velocity of the origin R(t).

The motion of the rigid body is governed by Newton's law:

$$\frac{d}{dt}P(t) = F(t)$$

$$\frac{d}{dt}L(t) = N(t)$$

with
$$P(t) = \rho^s \int_{G_2(t)} V(x,t)\,dx \qquad \text{linear momentum}$$

$$L(t) = \rho^s \int_{G_2(t)} x \times V(x,t)\,dx \qquad \text{angular momentum}$$

$$F(t), \; N(t) \qquad \qquad \text{total force and total torque about the origin O}$$

$$\rho^s \qquad \qquad \text{constant density of the rigid body}$$

For further reasoning, it is convenient to express the equations of motion in terms of the kinetic energy of the rigid body:

$$(2.1) \qquad \frac{d}{dt} \frac{\partial T^S}{\partial U} = F$$

$$(2.2) \qquad \frac{d}{dt} \frac{\partial T^S}{\partial \omega} + U \times \frac{\partial T^S}{\partial U} = N_R$$

with $\quad T^S = \frac{1}{2} \rho^S \int_{G_2(t)} V(x,t)^2 dx \quad$ the kinetic energy

$\qquad N_R \qquad\qquad\qquad\qquad$ the total torque about $R(t)$

The kinetic energy of the rigid body can be written in the form

$$T^S = \frac{1}{2} m^S U^2 + \frac{1}{2} \omega^T . I^S \omega + \omega^T . G^S U$$

with $\quad m^S \quad$ the mass of the rigid body

$\qquad I^S \quad$ the symmetric inertia tensor of the rigid body about R, whose components are given by

$$I^S_{ij} = \rho^S \int_{G_2} [\,(x_i - R_i)^2 \, \delta_{ij} - (x_i - R_i)(x_j - R_j)\,] \, dx$$

$(\delta_{ij}$ denotes the Kronecker symbol)

$$G^S = m^S \begin{pmatrix} O & S^S_3 & -S^S_2 \\ -S^S_3 & O & S^S_1 \\ S^S_2 & -S^S_1 & O \end{pmatrix} , \text{ an antisymmetric tensor,}$$

where $S^S = (S^S_1, S^S_2, S^S_3)$ is given by

$$S^S = \frac{1}{\text{vol}(G_2)} \int_{G_2} (x-R) dx \quad (\text{vol}(G_2) \text{ denotes the volume of } G_2.)$$

Note that $S^S = \bar{S}^S - R$ with $\bar{S}^S$ the center of mass of G_2.

2.2   Force and torque experienced by the rigid body

The total force and the total torque acting on the rigid body consist of two parts:

$$F = F^1 + F^{ex},$$

$$N_R = N_R^1 + N_R^{ex}.$$

F^1 and N^1 denote the total force and total torque, resp., caused by the surrounding liquid:

$$(2.3) \quad F^1 = \int_{\partial G_2} p . n \, dS$$

$$(2.4) \quad N_R^1 = \int_{\partial G_2} p[(x-R) \times n] \, dS$$

with ∂G_2 the surface of G_2

        p the pressure of the liquid

        n the inward normal unit vector with respect to the surface of G_2

F^{ex} and N_R^{ex} describe the action of further external forces, (such as gravitation). Assuming that a potential Ω per unit mass exists for these forces, we have

$$F^{ex} = -\rho^s \int_{G_2} \text{grad} \Omega \, dx = \rho^s \int_{\partial G_2} \Omega n \, dS,$$

$$N_R^{ex} = -\rho^s \int_{G_2} (x-R) \times \text{grad} \Omega \, dx = \rho^s \int_{\partial G_2} \Omega[(x-R) \times n] \, dS.$$

2.3   The equations of motion for the liquid

We assume that the liquid completely encloses the rigid body and is bounded by some fixed wall ∂G_1, the surface of a fixed domain $G_1 \subset \mathbb{R}^3$. The region occupied by the liquid at time t is denoted by $G(t) \subset \mathbb{R}^3$.

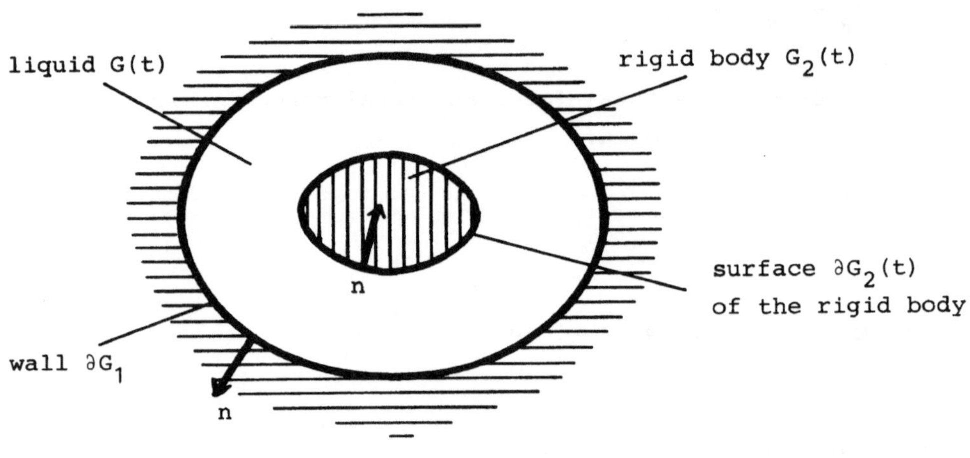

Fig. 2

The liquid is assumed to be ideal, i.e. incompressible and nonviscous. Then Euler's equations govern the motion of the liquid:

$$(2.5) \quad \rho^1 \cdot \left(\frac{\partial v_i}{\partial t} + v.\mathrm{grad}\ v_i \right) = - \frac{\partial p}{\partial x_i} - \rho^1 \cdot \frac{\partial \Omega}{\partial x_i} , \quad i = 1,2,3$$

$$(2.6) \quad \mathrm{div}\ v = 0$$

with ρ^1 the constant density of the liquid,

        $v = (v_1, v_2, v_3)$ the velocity field of the liquid.

The equations (2.5) describe the balance of momentum, equation (2.6) the conservation of mass.

Furthermore, we assume that the motion is irrotational, i.e. a function ϕ exists in G such that

(2.7) $v = -\text{grad}\phi$

Then (2.5) simplifies to

(2.8) $p = \rho^{1}.(\frac{\partial\phi}{\partial t} - \frac{1}{2}(\text{grad}\phi)^{2} - \Omega) + \text{constant}$

(2.6) and (2.7) lead to the Laplacian equation for ϕ:

(2.9) $\Delta\phi = 0$ in G

On the interface ∂G_2 the normal velocities of rigid body and liquid have to coincide and, on the fixed wall ∂G_1, the normal velocity of the liquid must vanish. This leads to the following boundary conditions:

(2.1o) $-\frac{\partial\phi}{\partial n} = n.V$ on ∂G_2

(2.11) $-\frac{\partial\phi}{\partial n} = 0$ on ∂G_1

where n denotes the outward normal unit vector with respect to G (this is consistent with the notation of Section 2.2)

Summary: The system rigid body-liquid is described by the equations of motion (2.1), (2.2) for the rigid body and (2.8), (2.9) for the liquid. The equations for the rigid body and the equations for the liquid are coupled by the boundary conditions (2.1o), (2.11) and the expressions (2.3), (2.4).

2.4   Transformation to the body system

For technical reasons the whole problem is expressed in terms of the coordinate system fixed in the rigid body. The transformation of a point x, a functional h and a vector function H is given by

$$x' = D(t)^T (x - R(t)),$$

$$h'(x',t) = h(x,t),$$

$$H'(x',t) = D(t)^T H(x,t),$$

where the prime indicates the corresponding representation in the body system. This transformation leads to the following system of equations in the body system (for simplicity the primes are dropped):

$$\frac{d}{dt} \frac{\partial T^S}{\partial U} + \omega \times \frac{\partial T^S}{\partial U} = F$$

$$\frac{d}{dt} \frac{\partial T^S}{\partial \omega} + \omega \times \frac{\partial T^S}{\partial \omega} + U \times \frac{\partial T^S}{\partial U} = N$$

with
$$T^S = \frac{1}{2} m^S U^2 + \frac{1}{2} \omega^T I^S \omega + \omega^T G^S U$$

$$I^S_{ij} = \rho^S \cdot \int_{G_2} (x_i^2 \delta_{ij} - x_i x_j) \, dx$$

$$G^S = m^S \begin{pmatrix} 0 & S_3^S & -S_2^S \\ -S_3^S & 0 & S_1^S \\ S_2^S & -S_1^S & 0 \end{pmatrix},$$

$$S^S = \frac{1}{\text{vol}(G_2)} \int_{G_2} x \, dx$$

and
$$F = F^{ex} + F^1, \quad N = N^{ex} + N^1$$

$$F^{ex} = \rho^S \int_{\partial G_2} \Omega n \, dS, \quad N^{ex} = \rho^S \int_{\partial G_2} \Omega (x \times n) \, dS$$

$$(2.12) \quad F^1 = \int_{\partial G_2} pn \, dS, \quad N^1 = \int_{\partial G_2} p(x \times n) \, dS$$

Furthermore,

$$(2.13) \quad p = \rho^1 \left(\frac{\partial \phi}{\partial t} - V.\mathrm{grad}\phi - \frac{1}{2} (\mathrm{grad}\phi)^2 - \Omega \right) + \text{constant}$$

$$\Delta\phi = 0 \qquad \text{in } G$$

$$(2.14) \quad - \frac{\partial \phi}{\partial n} = n.V \qquad \text{on } \partial G_2$$

$$- \frac{\partial \phi}{\partial n} = 0 \qquad \text{on } \partial G_1$$

with $\quad V = U + \omega \times x$

Note that G_2 is fixed in the body system, while ∂G_1 and G depend on t.

The aim of the two subsequent sections is to express F^1 and N^1 in terms of the kinetic energy of the liquid.

2.5 The kinetic energy of the liquid

The velocity field of the liquid is determined by the boundary value problem (2.14). We may write, after Kirchhoff,

$$\phi = U_1\phi_1 + U_2\phi_2 + U_3\phi_3 + \omega_1\chi_1 + \omega_2\chi_2 + \omega_3\chi_3.$$

Then ϕ solves (2.14) if, for i = 1,2,3,

$$\Delta\phi_i = 0 \qquad \text{in } G$$

$$(2.15) \quad - \frac{\partial \phi_i}{\partial n} = n_i \qquad \text{on } \partial G_2$$

$$\frac{\partial \phi_i}{\partial n} = 0 \qquad \text{on } \partial G_1$$

and $\quad \Delta\chi_i = 0 \qquad \text{in } G$

$$- \frac{\partial \chi_i}{\partial n} = (x \times n)_i \quad \text{on } \partial G_2$$

(2.16)

$$\frac{\partial \chi_i}{\partial n} = 0 \qquad \text{on } \partial G_1$$

For the kinetic energy T^1 of the liquid we obtain

$$T^1 = \frac{1}{2} \rho^1 \int_G v^2 \, dx = \frac{1}{2} \rho^1 \int_G (\text{grad}\phi)^2 dx =$$

$$= \frac{1}{2} U^T M^1 U + \frac{1}{2} \omega^T I^1 \omega + \omega^T G^1 U$$

with symmetric tensors M^1 and I^1, whose components are

$$M^1_{ij} = \rho^1 \cdot \int_G \text{grad}\phi_i \cdot \text{grad}\phi_j \, dx$$

$$I^1_{ij} = \rho^1 \cdot \int_G \text{grad}\chi_i \cdot \text{grad}\chi_j \, dx,$$

and a coupling tensor G^1, given by

$$G^1_{ij} = \rho^1 \cdot \int_G \text{grad}\phi_i \cdot \text{grad}\chi_j \, dx.$$

The components M^1_{ij}, I^1_{ij}, G^1_{ij} are called <u>hydrodynamic coefficients</u>. In general, there are 21 different coefficients.

2.6   <u>The action of the liquid on the rigid body</u>

From (2.12) and (2.13), it follows by applying Gauss' Theorem and Stokes' Theorem that

$$(2.17) \quad F^1 = \frac{d}{dt} \left(\rho^1 \int_{\partial G_2} \phi n \, dS \right) + \omega \times \left(\rho^1 \cdot \int_{\partial G_2} \phi n \, dS \right) - \rho^1 \cdot \int_{\partial G_2} \Omega n \, dS +$$

$$+ \frac{1}{2} \rho^1 \int_{\partial G_1} v^2 n \, dS$$

$$(2.18) \quad N^1 = \frac{d}{dt} \left(\rho^1 \int_{\partial G_2} \phi (x \times n) dS \right) + \omega \times \left(\rho^1 \int_{\partial G_2} \phi (x \times n) dS \right) +$$

$$+ U \times \left(\rho^1 \int_{\partial G_2} \phi (x \times n) dS \right) - \rho^1 \int_{\partial G_2} \Omega (x \times n) dS +$$

$$+ \frac{1}{2} \rho^1 \int_{\partial G_1} v^2 (x \times n) dS$$

We assume that v almost vanishes at the boundary ∂G_1. This seems to be justified if the body G_2 performs small vibrations in the liquid that was originally at rest. Then we can neglect the last expression in (2.17) and (2.18), respectively. Furthermore, the integral expressions in (2.17) and (2.18) involving the velocity potential ϕ can be written in terms of the kinetic energy by using Gauss' Theorem:

$$\rho^1 \int_{\partial G_2} \phi n \, dS = - \frac{\partial T^1}{\partial U}$$

$$\rho^1 \int_{\partial G_2} \phi (x \times n) \, dS = - \frac{\partial T^1}{\partial \omega}$$

So, finally, we have

$$F^1 = - \frac{d}{dt} \frac{\partial T^1}{\partial U} - \omega \times \frac{\partial T^1}{\partial U} - \rho^1 \int_{\partial G_2} \Omega n \, dS$$

$$N^1 = - \frac{d}{dt} \frac{\partial T^1}{\partial \omega} - \omega \times \frac{\partial T^1}{\partial \omega} - U \times \frac{\partial T^1}{\partial U} - \rho^1 \int_{\partial G_2} \Omega (x \times n) \, dS$$

Using this representation for F^1 and N^1 we obtain for the equations of motion of the rigid body:

$$\frac{d}{dt} \frac{\partial T}{\partial U} + \omega \times \frac{\partial T}{\partial U} = \left(1 - \frac{\rho^1}{\rho^s}\right) . F^{ex}$$

$$\frac{d}{dt} \frac{\partial T}{\partial \omega} + \omega \times \frac{\partial T}{\partial \omega} + U \times \frac{\partial T}{\partial U} = (1 - \frac{\rho^1}{\rho^s}) \cdot N^{ex}$$

with $T = T^s + T^1 = \frac{1}{2} U^T M U + \frac{1}{2} \omega^T I \omega + \omega^T G U$

$$M = m^s \cdot E + M^1, \quad I = I^s + I^1, \quad G = G^s + G^1$$

(E identity matrix)

These equations differ from the corresponding equations of motion of the rigid body in air (better in vacuum)

a. by a reduced action of the external force (buoyancy) and

b. by a virtually increased kinetic energy. For a simple one-dimensional translational motion (say in x-direction), we have

$$T = \frac{1}{2} (m^s + M^1_{11}) U^2_1 \text{ with } M^1_{11} = \rho^1 \int_G (grad\phi_1)^2 dx > 0,$$

which means that the effect of the surrounding liquid leads to more inertia, expressed by a virtually increased mass, M^1_{11} is the hydrodynamic mass for the x-direction.

Note that, in general, m^s, I^s and G^s are constant (in the body system), while M, I and G depend on the location of the body G_2 relative to the wall ∂G_1.

2.7   <u>Hydrodynamic coefficients for bodies of revolution</u>

If ∂G_1 and G_2 are symmetric about the same axis, say the x_3-axis, the calculation of the hydrodynamic coefficients simplifies considerably. Only 4 non-trivial coefficients remain:

$$M^1 = \begin{pmatrix} M^1_{11} & 0 & \\ 0 & M^1_{11} & 0 \\ 0 & 0 & M_{33} \end{pmatrix}, \quad I^1 = \begin{pmatrix} I^1_{11} & & 0 \\ 0 & I^1_{11} & 0 \\ 0 & 0 & 0 \end{pmatrix}, \quad G^1 = \begin{pmatrix} 0 & G^1_{12} & 0 \\ -G_{21} & 0 & 0 \\ 0 & 0 & 0 \end{pmatrix}$$

The inertia tensor I^s and the coupling tensor G^s for the rigid body are of the form

$$I^S = \begin{pmatrix} I^S_{11} & 0 & 0 \\ 0 & I^S_{11} & 0 \\ 0 & 0 & I^S_{33} \end{pmatrix}, \quad G^S = m^S \begin{pmatrix} 0 & S^S_3 & 0 \\ -S^S_3 & 0 & 0 \\ 0 & 0 & 0 \end{pmatrix}$$

Then the kinetic energy $T = T^S + T^l$ simplifies to

$$T = \tfrac{1}{2}\,(m^S + M^l_{11})[U^2_1 + U^2_2] + \tfrac{1}{2}\,(m^S + M^l_{33})U^2 +$$

$$\tfrac{1}{2}\,(I^S + I^l_{11})[\omega^2_1 + \omega^2_2] + \tfrac{1}{2}\,I^S_{33}\omega^2_3$$

if

(2.19) $\quad m^S \cdot S^S_3 + G^l_{12} = 0 \quad$ in the body system.

This last condition is satisfied if the origin R of the body system is defined as

$$R = (\bar{S}^S_1, \bar{S}^S_2, \bar{S}^S_3 + \tfrac{1}{m^S}\,G_{12}) \quad \text{in the space system.}$$

with $\bar{S}^S = (\bar{S}^S_1, \bar{S}^S_2, \bar{S}^S_3)$ the center of mass of G_2.

<u>Summary:</u> The body submerged in liquid behaves as though

a. its mass is increased by M^l_{11} for a translational motion in a plane normal to the x_3-axis resp. by M^l_{33} for a translational motion parallel to the x_3-axis,

b. its moment of inertia about an axis normal to the x_3-axis is increased by I^l_{33},

c. its x_3-coordinate of the center of mass is translated by G_{12}/m^S,

compared to the body without surrounding liquid. The coefficients are given by

$$M_{11}^1 = M_{22}^1 = \alpha_r m^1 \qquad \alpha_r = \frac{1}{\mathrm{vol}(G_2)} \int_G (\mathrm{grad}\phi_1)^2 dx$$

$$M_{33}^1 = \alpha_z m^1, \qquad \alpha_z = \frac{1}{\mathrm{vol}(G_2)} \int_G (\mathrm{grad}\phi_3)^2 dx$$

$$I_{11}^1 = I_{22}^1 = \beta \cdot i^1, \quad \beta = \frac{1}{\mathrm{mom}(G_2)} \int_G (\mathrm{grad}\chi_2)^2 dx$$

$$G_{12} = \gamma \cdot m^1, \qquad \gamma = \frac{1}{\mathrm{vol}(G_2)} \int_G \mathrm{grad}\phi_1 \cdot \mathrm{grad}\chi_2 \, dx$$

with $m^1 = \rho^1 \cdot \mathrm{vol}(G_2)$, $i^1 = \rho^1 \cdot \mathrm{mom}(G_2)$, $\mathrm{mom}(G_2) = \int_{G_2} (x_2^2 + x_3^2) \, dx$. All other hydrodynamic coefficients vanish.

The potentials ϕ_1, ϕ_3 and χ_2 are the solutions of the following boundary value problems:

Problem A:
$$\Delta\phi_1 = 0 \qquad \text{in } G$$
$$-\frac{\partial\phi_1}{\partial n} = n_1 \qquad \text{on } \partial G_2$$
$$\frac{\partial\phi_1}{\partial n} = 0 \qquad \text{on } \partial G_1$$

Problem B:
$$\Delta\chi_2 = 0 \qquad \text{in } G$$
$$-\frac{\partial\chi_2}{\partial n} = x_3 n_1 - x_1 n_3 \qquad \text{on } \partial G_2$$
$$\frac{\partial\chi_2}{\partial n} = 0 \qquad \text{on } \partial G_1$$

Problem C:
$$\Delta\phi_3 = 0 \qquad \text{in } G$$
$$-\frac{\partial\phi_3}{\partial n} = n_3 \qquad \text{on } \partial G_2$$
$$\frac{\partial\phi_3}{\partial n} = 0 \qquad \text{on } \partial G_1$$

Note that the factors α_r, α_z, β and γ depend only on the geometry of the problem.

3   <u>The mathematical model</u>

The calculation of the hydrodynamic coefficients leads to the solution of several von Neumann boundary value problems of the form

$$(3.1) \quad \Delta u = 0 \qquad \text{in } G$$

$$\frac{\partial u}{\partial n} = g \qquad \text{on } \partial G$$

We assume that G is a bounded, open and simply connected subset of $\mathbb{R}^3$, whose boundary ∂G is Lipschitzian (i.e. the boundary is locally represented by Lipschitz continuous functions).

A solution to (3.1) also solves the variational problem

$$(3.2) \quad a(u,v) = l(v) \quad \text{for each } v \in E$$

with

$$a(u,v) = \int_G \text{grad } u \cdot \text{grad } v \, dx$$

$$l(v) = \int_{\partial G} g.v \, dS$$

$$E = \{v \in H^1(G) \mid \int_G v \, dx = 0\}$$

where $H^1(G)$ denotes the Sobolev space of all functions on G with square-integrable (weak) first derivatives. A solution in E to (3.2) is called a <u>weak solution</u> of the von Neumann boundary value problem.

The famous Lax-Milgram lemma, see [2], guarantees the existence of a weak solution:

Theorem 3.1: Assume that $g \in L_2(\partial G)$ (i.e. g is square-integrable on ∂G). Then

a. a weak solution in E of the von Neumann boundary value problem exists if and only if

$$(3.3) \qquad \int_{\partial G} g \, dS = 0,$$

b. the solution is unique in E;

c. the solution depends continuously on the data g.

Proof: see [3]

Using Gauss' Theorem one easily shows that condition (3.3) is satisfied for each of the boundary value problems (2.15) resp. (2.16). Therefore, the problem of calculating the potentials ϕ_i and χ_i, $i = 1,2,3$, is well posed: A unique solution exists, which continuously depends on the data. The hydrodynamic coefficients can be expressed by the form $a(u,v)$ (up to some physical constants):

$$M_{ij}^1 = \rho^1 \cdot a(\phi_i, \phi_j)$$

$$I_{ij}^1 = \rho^1 \cdot a(\chi_i, \chi_j)$$

$$G_{ij}^1 = \rho^1 \cdot a(\phi_i, \chi_j).$$

Next, we give a well-known characterization of the solution to the variational problem (3.2):

Theorem 3.2: Assume the notations and hypotheses of Theorem 3.1. The solution to (3.2) is the unique minimizer of the form

$$f(u) = \frac{1}{2} a(u,u) - l(u).$$

Proof: see [3].

More can be said for bodies of revolution. The mapping

$$Z: \; (0,\infty) \times \mathbb{R} \times [0,2\pi[\; \to \; \mathbb{R}^3$$

$$(r,z,\varphi) \to (r\cos\varphi, \; r\sin\varphi, z) = (x_1,x_2,x_3)$$

describes the transformation from cylindrical coordinates to
Cartesian coordinates. If G is symmetric about the x_3-axis, it
can be written in the form

$$G = Z(S \times [0,2\pi[) \text{ with } S \subset (0,\infty) \times \mathbb{R}.$$

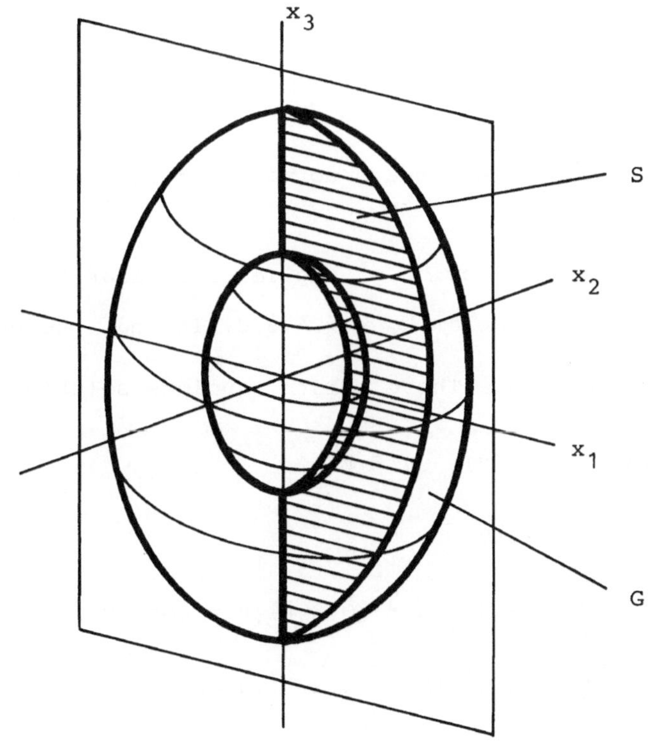

Fig. 3

The space $H^1(G)$ can be identified by the corresponding space
$H^1_{2\pi}(S \times [0,2\pi[)$ of functions in cylindrical coordinates. The
weak form of Problem k, k = A,B,C, in cylindrical coordinates is

$$a(u,v) = 1^k(v)$$

where
$$a(u,v) = \int_S \int_0^{2\pi} \left(\frac{\partial u}{\partial r} \frac{\partial v}{\partial r} + \frac{\partial u}{\partial z} \frac{\partial v}{\partial z} + \frac{1}{r^2} \frac{\partial u}{\partial \varphi} \frac{\partial v}{\partial \varphi} \right) r\, d\varphi\, dr\, dz$$

and
$$1^k(v) = \int_{\partial S} \int_0^{2\pi} g^k \cdot v \cdot r\, d\varphi\, ds$$

with
$$g^k = \begin{cases} -n_r \cdot \cos\varphi & k = A \\[2mm] (r \cdot n_z - z n_r) \cos\varphi & \text{for } k = B \quad \text{on } \partial G_2 \\[2mm] -n_z & k = C \\[2mm] 0 & \text{on } \partial G_1 \end{cases}$$

(n_r, n_z) denotes the outward normal unit vector with respect to the boundary of S.

The special form of g^k implies a similar form for the solutions u^k of Problem k, k = A,B,C. Indeed, one easily shows

Theorem 3.3: There exists a unique solution $u^k \in E^k$ to

$$a(u,v) = 1^k(v), \quad v \in E^k$$

for k = A, B, C with

$$E^k = \begin{cases} \{u \in H^1_{2\pi}(S \times [0,2\pi[) \,|\, u = \cos\varphi \cdot v(r,z)\} & k = A,B \\[3mm] \{u \in H^1_2(S \times [0,2[) \,|\, u = v(r,z),\ \int_S v \cdot r\, dr\, dz = 0\} & k = C \end{cases}$$

Proof: see [3].

4   The numerical method

The Ritz method is used to obtain an approximate solution
to the variational problem (3.2): The Hilbert space E is replaced by a finite-dimensional space E_h with basis $\{\psi_1,\ldots,\psi_N\}$,
the form f on E is approximated by a form

$$f_h(u) = \frac{1}{2} a_h(u,u) - l_h(u)$$

on E_h with a_h a positive definite bilinear form and l_h a linear
functional. The unique minimizer u_h of f_n in E_h is chosen as
approximate solution to (3.2). For the actual computation of u_h,
we set

$$u_h = \sum_{j=1}^{N} \alpha_j \psi_j.$$

Then we have: u_h minimizes f_h if and only if

$$K.\alpha = b$$

with $K = (K_{ij})$, $K_{ij} = a_h(\psi_i,\psi_j)$, $b = (b_i)$, $b_i = l_h(\psi)$ and
$\alpha = (\alpha_j)$. Hence, the problem reduces to the solution of a linear
system with positive definite matrix K. This is done by a band
Cholesky decomposition.

It remains to specify the space E_h with an appropriate
basis and the forms a_h resp. l_h on E_h for each of the problems
A, B and C.

4.1   The finite-dimensional spaces E_h

The finite element technique is used to construct E_h:
First, the cross section S in the (r,z)-plane is triangulated.
The triangles build up a polygonal approximation S_h of S, see
Fig. 4.

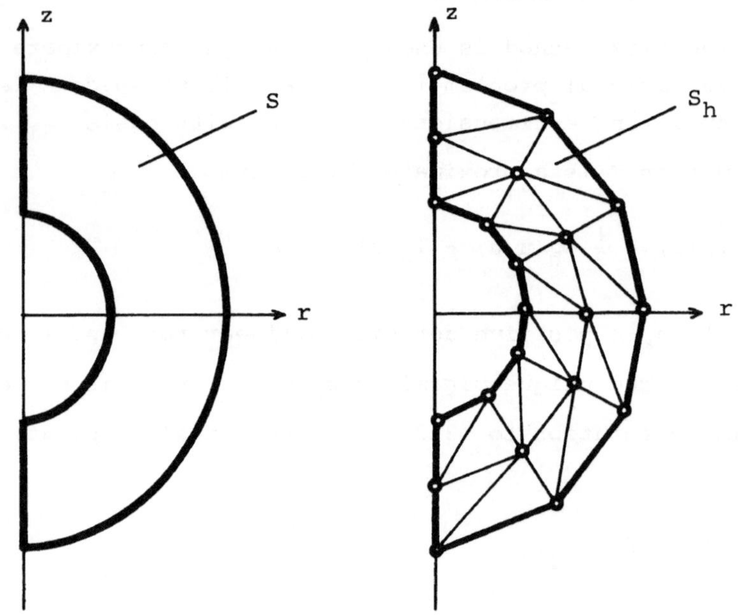

Fig. 4

Let $PL(S_h)$ be the set of piecewise linear and continuous

functions on the triangulation. The vertices of the triangles

are denoted by P_i, $i = 1,\ldots,M$. To each vertex P_i there is an

associated basis function $p_i \in PL(S_h)$ with

$$p_i(P_j) = \delta_{ij}$$

These basis functions form a basis for the space $PL(S_h)$.

Theorem 3.3 suggests the following choice for E_h^k,
$k = A,B,C.$

$$E_h^k = \begin{cases} \{u = \cos\varphi\, p(r,z) \mid p \in PL(S_h), p(0,z) = 0\} & k = A,B \\ & \text{for} \\ \{u = p(r,z) \mid p \in PL(S_h), \int_{S_h} p \cdot r\,dr\,dz = 0\} & k = C \end{cases}$$

Remark: The condition $p(0,z) = 0$ guarantees that $E_h^k \subset H_2^1 (S_h \times [0,2\pi[)$ for $k = A,B$.

The basis functions of $PL(S_h)$ naturally generate a basis for E_h^k. The additional requirement $p(0,z)$ for $k = A,B$ excludes basis functions whose associated vertex lies on the z-axis. Instead of the scaling condition $\int_{S_h} p\, r\, drdz$ for $k = C$ one of the basis functions is excluded from the basis. This alters the solution by a constant, which does not effect the calculation of the hydrodynamic coefficients.

4.2 The forms a_h and l_h

Naturally, we set

$$a_h(u,v) = \int_{S_h} \int_0^{2\pi} (\frac{\partial u}{\partial r} \frac{\partial v}{\partial r} + \frac{\partial u}{\partial z} \frac{\partial v}{\partial z} + \frac{1}{r^2} \frac{\partial u}{\partial \varphi} \frac{\partial v}{\partial \varphi})\, r\, d\varphi drdz$$

$$= \begin{cases} \pi\int_{S_h} (\frac{\partial p}{\partial r} \frac{\partial q}{\partial r} + \frac{\partial p}{\partial z} \frac{\partial q}{\partial r} + \frac{1}{r^2} pq)\, rdrdz \\ \qquad\qquad\qquad\qquad\qquad \text{for } k = A,B \\ (u = \cos\varphi.p(r,z),\ v = \cos\varphi.q(r,z)) \\ \\ 2\pi\int_{S_h} (\frac{\partial p}{\partial r} \frac{\partial q}{\partial r} + \frac{\partial p}{\partial z} \frac{\partial q}{\partial z})\, rdrdz \\ \qquad\qquad\qquad\qquad\qquad \text{for } k = C \\ (u = p(r,z),\ v = q(r,z)) \end{cases}$$

and (notation as before)

$$l_h^k(v) = \int_{\partial S_h} \int_0^{2\pi} g^k.v.r\, d\varphi ds$$

$$= \begin{cases} -\pi \displaystyle\int_{\partial S_h} p.n_r.r \, ds & \text{for } k = A \\[2em] \pi \displaystyle\int_{\partial S_h} p.(rn_z - zn_r)r \, ds & \text{for } k = B \\[2em] -\pi \displaystyle\int_{\partial S_h} p.n_z.r \, ds & \text{for } k = C \end{cases}$$

<u>Remark:</u> The computation of $a_h(\psi_i, \psi_j)$ can be done explicitly. However, the resulting formulas are rather complicated. Therefore, a 7-point quadrature formula on a triangle was used to approximate $a_h(\psi_i, \psi_j)$, see [3].

4.3 <u>Convergence results</u>

Convergence can not be analysed in the space E because, in general, E_h is not a subset of E. Therefore, both E and E_h are imbedded into a global Hilbert space $\hat{E}$ in order to compare the approximation $u_h \in E_h$ with the exact solution $u \in E$.

In Cartesian coordinates we have

$$E^k \subset H^1(G) \quad \text{and} \quad E_h^k \subset H^1(G_h) \quad \text{for } k = A,B,C$$

with $G_h = Z(S_h \times [0,2\pi[)$ the Cartesian representation of the domain $S_h \times [0,2\pi[$. Let $\hat{G}$ be a bounded open subset of $\mathbb{R}^3$ with

$$G \subset \hat{G}, \quad G_h \subset \hat{G}.$$

Then, for

$$\hat{E} = L_2(\hat{G})^4 = L_2(\hat{G}) \times L_2(\hat{G}) \times L_2(\hat{G}) \times L_2(\hat{G})$$

we have $E^k \subset \hat{E}$ and $E_h^k \subset \hat{E}$ for $k = A,B,C$ by means of the imbeddings

$$J:\ H^1(G) \to L_2(\hat{G})^4,\ J(u) = \begin{cases} (u,\ \dfrac{\partial u}{\partial x_1},\ \dfrac{\partial u}{\partial x_2},\ \dfrac{\partial u}{\partial x_3}) & \text{on } G \\[2ex] 0 & \text{else} \end{cases}$$

$$J_h:\ H^1(G_h) \to L_2(\hat{G})^4,\ J_h(u) = \begin{cases} (u,\ \dfrac{\partial u}{\partial x_1},\ \dfrac{\partial u}{\partial x_2},\ \dfrac{\partial u}{\partial x_3}) & \text{on } G_h \\[2ex] 0 & \text{else} \end{cases}$$

One can show that the approximate solution $u_h^k \in E_h^k \subset \hat{E}$ converges towards the exact solution $u^k \in E^k \subset \hat{E}$ of Problem k, $k = A,B,C$ provided

a. the mesh size of the triangulation tends to zero,

b. S_h converges to S as the mesh size tends to zero, and

c. the smallest angle of the triangles as well as the smallest angle between the z-axis and a triangle with a vertex on the z-axis are bounded from below as the mesh size tends to zero.

Moreover, if $G_h = G$, we have the following monotony result:

$$a_h(u_h^k, u_h^k) \leq a(u^k, u^k),\quad k = A,B,C,$$

which means that the approximations for the hydrodynamic co-efficients M_{11}^1, M_{33}^1 and I_{11}^1 are lower bounds.

For details and proofs of these convergence results, see [3].

4.4 Numerical experiments

First, the numerical method is tested on a problem with known exact solution. Then a model of a turbine in a pipe filled with water is considered. All computations were performed on an IBM 360/44 system in single precision.

<u>Example A:</u> ellipsoids of revolution

The surfaces ∂G_2 and ∂G_1 are confocal ellipsoids of revolution.
The common foci are $(0,0,1)$ and $(0,0,-1)$. The semiminor axis are
1 and 2, resp., see Fig. 5.

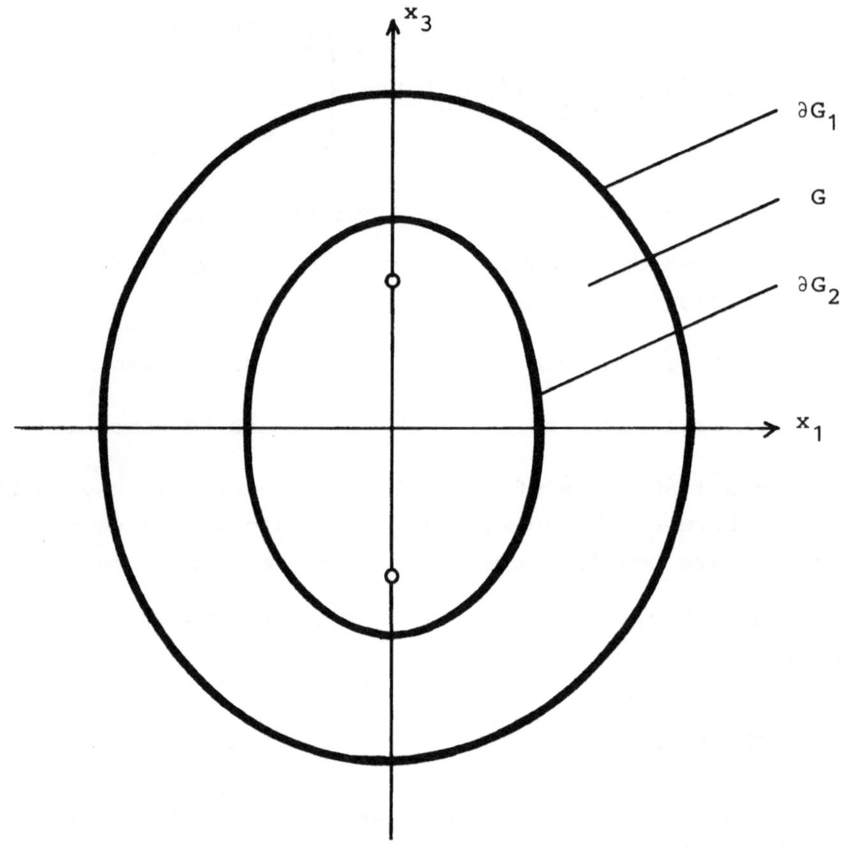

Fig. 5

The inner and outer boundary of S are uniformly divided into
MZ parts, the number of subdivisions between inner and outer
boundary is MR. In Fig. 6 a typical triangulation iwth MR = 3
and MZ = 12 is shown.

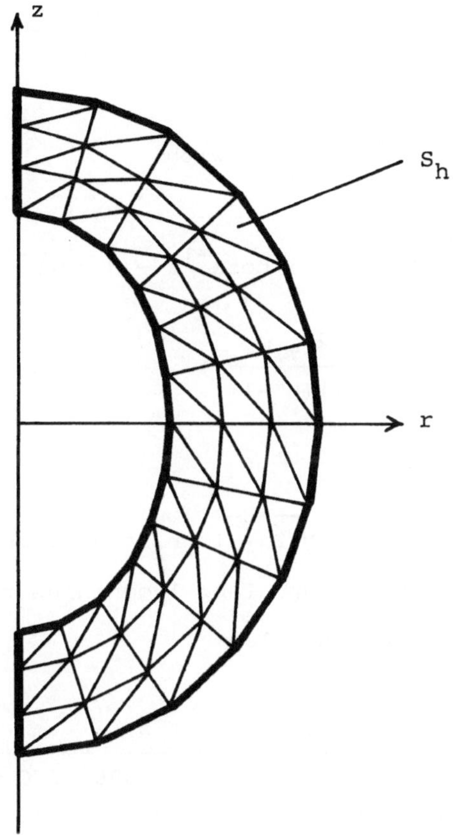

Fig. 6

The hydrodynamic coefficients for this problem are explicitly known. By means of ellipsoidal harmonics one obtains

$$\alpha_r = \frac{-2B-a(a^2-1)}{a(a^2-1)} \qquad \text{with}$$

$$\frac{1}{B} = \coth^{-1}b - \coth^{-1}a + \frac{a^2-2}{a(a^2-1)} - \frac{b^2-2}{b(b^2-1)}$$

$$\alpha_z = \frac{B-a(a^2-1)}{a(a^2-1)} \qquad \text{with}$$

$$\frac{1}{B} = \coth^{-1}b - \coth^{-1}a + \frac{a}{a^2-1} - \frac{b}{b^2-1}$$

$$\beta = \frac{2B-a(a^2-1)}{a(a^2-1)(2a^2-1)^2} \quad \text{with}$$

$$\frac{1}{B} = (2a^2-1)(3 \coth^{-1}a - 3 \coth^{-1}b - \frac{a(6a^2-7)}{(2a^2-1)(a^2-1)}$$

$$+ \frac{b(6b^2-7)}{(2b^2-1)(b^2-1)})$$

$$\gamma = 0$$

where $a = \sqrt{r_2^2 + 1}$, $b = \sqrt{r_2^2 + 1}$, r_2 and r_1 being the semiminor axes of ∂G_2 and ∂G_1, resp. Tab. 1 contains the obtained numerical results for MZ = 50, MR = 10:

	α_r	α_z	β	γ
exact	0.92260	0.55396	0.08141	0.0
numerical	0.91877	0.55006	0.08056	0.00006
error	0.42 %	0.70 %	1.04 %	–

Tab. 1

As Tab. 1 shows, the numerical approximations are in good agreement with the exact solutions. Moreover, the numerical solutions are lower bounds (as expected, although G and G_h slightly differ).

Example B: turbine in a pipe
Fig. 7 gives a simple model of a turbine. The turbine is placed in the middle of the pipe with respect to the x_1- and x_3-direction. MZ+1 grid points are (almost) uniformly distributed along the inner boundary, thus dividing the boundary into MZ segments.

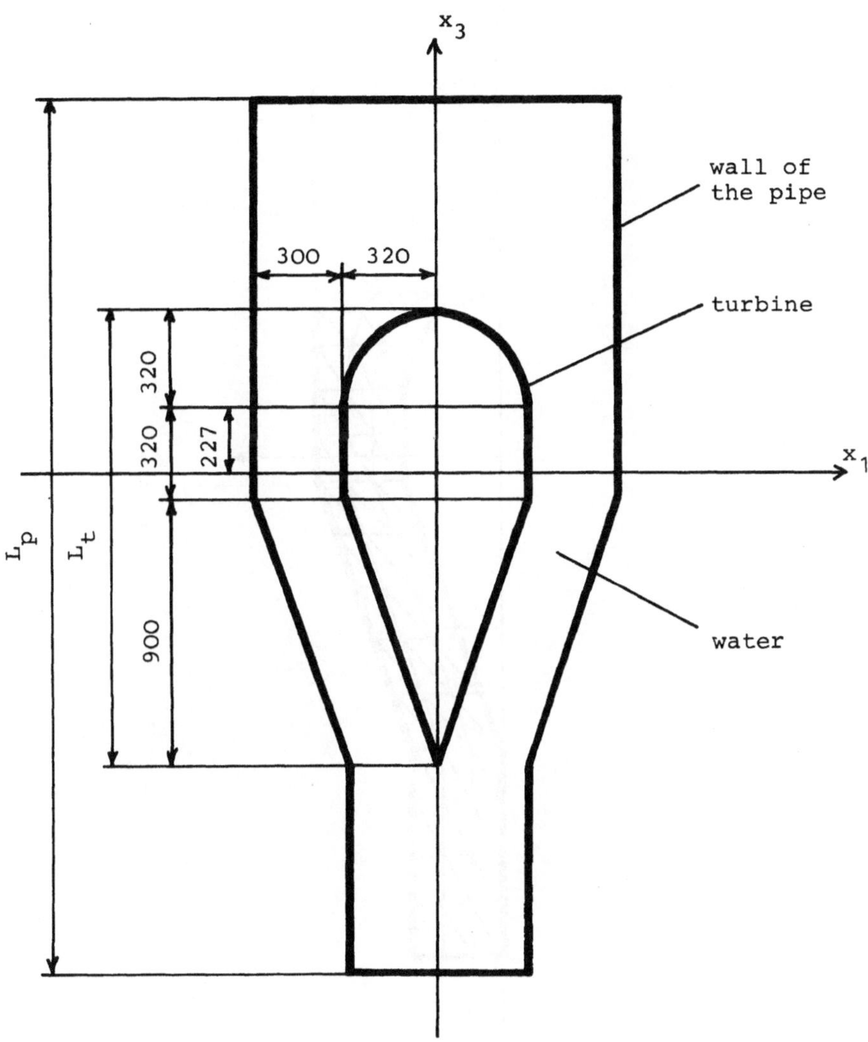

Fig. 7

The distance in x_1-direction to the outer boundary is divided into MZ parts. A typical element grid is shown in Fig. 8:

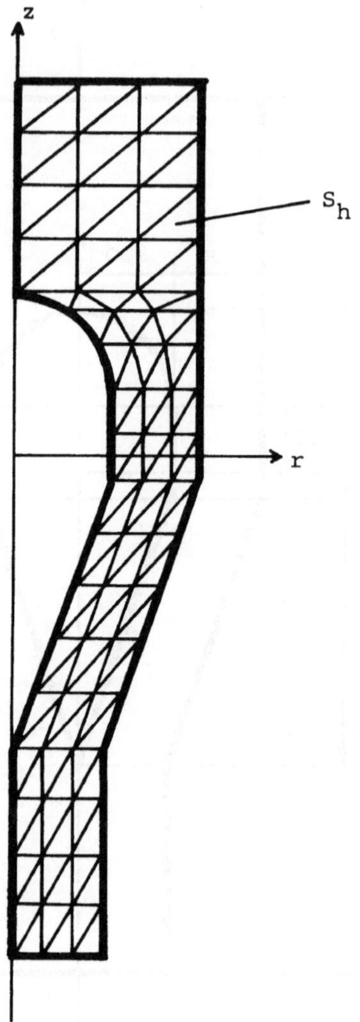

Fig. 8

Naturally, the solution depends on the total length L_p of the pipe. In order to analyze the effect of the length L_p on the solution, the problem was solved for

$$L_p = i\, L_t, \quad i = 1,2,3,4,5$$

with L_t the length of the turbine.

Rule of thumb for the choice of MR and MZ:
Given a total number $K = 2MR.MZ$ of triangle, here $K \sim 500$. Then, MR and MZ are chosen such that

$$f_r/f_z \sim 0.8$$

where $f_r = \dfrac{F}{MR.L_p}$ (F is the area of S) and $f_z = \dfrac{L_p}{MZ}$. (f_r and f_z roughly measure how fine the grid is in r-direction and in z-direction, resp. A value of about 0.8 for the ratio f_r/f_z led to rather good results for various test problems.)

Tab. 2 shows the numerical results:

	MZ/MR	α_r	α_z	β	γ
i=1	36/14	1.25701	0.58019	0.40098	-0.07459
i=2	56/9	1.02779	0.46789	0.35999	-0.32426
i=3	62/8	1.02375	0.46517	0.35768	-0.33148
i=4	71/7	1.02143	0.46285	0.35551	-0.33148
i=5	83/6	1.01973	0.46117	0.35397	-0.33115

Tab. 2

The values remain about constant if L_p exceeds $2L_t$.

Acknowledgements.

The author is very grateful to Dr.D.Fischer and Dr.H.Steiner (both VOEST-ALPINE AG) for suggesting the subject and for the interesting discussions, and to Prof.Dr.Hj.Wacker for the valuable advice.

References:

[1] Lamb, H.: Hydrodynamics. 6th ed. London: Cambridge University Press 1932.

[2] Strang, G.; Fix, G.J.: An Analysis of the Finite Element Method. Englewood Cliffs: Prentice Hall 1973.

[3] Zulehner, W.: Über die Berechnung der hydrodynamischen Koeffizienten für den rotationssymmetrischen Fall. Diploma thesis, University of Linz 1978.

Project 3

CONTROL OF THE SOLIDIFICATION FRONT BY SECONDARY COOLING IN CONTINUOUS CASTING OF STEEL

Heinz W.Engl and Thomas Langthaler

1 The Problem

In the production of rolled steel, there are two possible process routes between steel making and hot rolling. The older process is "block casting", the newer one is "continuous casting". The latter process has many advantages over the former: First, the resulting product is already in a form usable for the final rolling process, an intermediate rolling process is saved. Ideally, it can be rolled directly after leaving the casting machine ("direct rolling"), which saves the enormous amounts of energy used in the pusher furnace for reheating the steel slabs (cf. the article by Auzinger and Wacker in this volume). Also, in block casting, some of the material at the top and at the bottom of the blocks has to be cut off because of impurities. The relation between usable cast steel and used liquid steel is about 10 % better in continuous casting compared to block casting. In the VOEST-Alpine steel company in Linz, for which we carried out the project to be described here, about 90 % of the steel is processed by continuous casting.

Our working groups in Linz are involved with the process of steel making in various ways: Besides this project and the project concerning the pusher furnace mentioned above, we are also involved with the problem of scheduling the different processes in steel making, continuous casting and direct hot rolling in such a way that a given "product mix" of cast steel of different width is produced in a given period of time (cf. [13]).

In continuous casting, products of different cross-sections can be produced. Here, we are concerned with casting of

"slabs", whose cross-section is a rectangle of dimension
0.8 - 1.6 m by 0.2 - 0.3 m, in a "bow type" casting machine as
shown in Fig. 1. There, the liquid steel is cast into the
"mould" (upper left-hand corner) , from where it flows continuously into the casting machine. Initially, the strand must already be supported by a solid shell, although it still has a liquid core. This is necessary for pulling the strand through the
casting machine by rolls at a speed of around 1.5 m/min. In the
casting machine, heat must be extracted from the strand in the
"secondary cooling zones" by spraying water onto the strand. In
order to minimize defects, it is of utmost importance to control e.g. the surface temperature, its gradient and / or the
"front of solidification" (i. e., the boundary between solid
and liquid phase, see Section 2 for details) in an appropriate
way. In this project, the aim was to control the solidification
front via the spray cooling. Note that also the casting speed
influences the solidification front (even much more than the
spray cooling). However, in practice the casting speed cannot
easily be used as a control variable, since it depends heavily
on other parts of the steel making process. In practice, one
also wants to obtain the same solidification front as a function of position in the casting machine for different speeds,
which can also be achieved by the method to be described here.

For other practical aspects of our project cf. [8] and
[21]. In [31], it is described how the results of this project
are actually used in the VOEST-Alpine plant. Theoretical aspects motivated by our project are treated in [9] and [12].

2   The Mathematical Model and its Properties

Usually, phase change problems like ours are modelled as
Stefan problems . There, a heat equation with usually different
coefficient functions is valid in two regions (representing the
solid and the liquid phases), which are separated by a free
boundary, the solidification front. From the vast literature
about Stefan problems, we quote [14], [15], [4], [19], [30],
[3], [5].

In our problem, the soldification does not take place abruptly at a specific temperature. In the qualities of steels usually cast, solidification takes place within a temperature range between about 1460^0C and 1490^0C. This also implies that we will have a "mushy region", where both phases are mixed. Note that such a mushy region can also appear in (multi-dimensional) Stefan problems. Since there is no unique "solidification front", we have to decide which isotherm we actually want to control by appropriate spray cooling. It seems to be most important to control the boundary between solid phase and mushy region, for which we will use the term "solidification front" below. Thus, we are controlling the isotherm at about 1460^0 (depending on the specific quality of steel).

Because of the fact that solidification takes place over a relatively large temperature interval, it is natural to use the enthalpy for modelling our problem. If we denote by ρ the density of the material, by u its temparature, by $L = L(u)$ the latent heat content, and by $\tilde{c} = \tilde{c}(u)$ the specific heat at temperature u, then the enthalpy $H = H(u)$ can be defined by

$$(2.1) \quad H(u) = \int_{u_0}^{u} \rho\tilde{c}(T)\,dT + \rho L(u),$$

where u_0 is some reference temperature. If $[u_1, u_2]$ is the temperature range within which solidification takes place, then L is a monotone function with

$$(2.2) \quad L(u) = \begin{cases} 0 & \text{for } u < u_1 \\ L(u_2) & \text{for } u \geq u_2. \end{cases}$$

Using the enthalpy, heat conduction can be modelled by the equation

$$(2.3) \quad \frac{\partial H(u)}{\partial t} = \mathrm{div}(k(u).\mathrm{grad}\,u),$$

where $k = k(u)$ denotes the thermal conductivity at temperature u, t denotes time and div and grad are taken with respect to the space variables. Note that (2.3) is valid in all phases, and also across the phase boundaries.

If L is differentiable and $\tilde{c}$ is continuous, H is differentiable. Thus, $\frac{\partial H(u)}{\partial t} = \frac{dH}{du} \cdot \frac{\partial u}{\partial t}$, so that (2.3) can be written as

(2.4) $\quad \rho c(u) \cdot \frac{\partial u}{\partial t} = \mathrm{div}(k(u) \cdot \mathrm{grad}\ u)$
with
(2.5) $\quad c(u) := \frac{1}{\rho} \cdot \frac{dH}{du}(u)$.

c is called the "effective heat capacity".

Fig. 2 shows typical pictures of the enthalpy and of the (effective) heat capacity, the thermal conductivity and the density of steel as functions of temperature (cf. [17], [22]). Note that the density does not change much with the temperature, which justifies to assume a constant density.

We now turn to the model of continuous (slab) casting we use. By D , we denote the two-dimensional cross-section of the strand in (x,y)-space. It is known ([27]) that heat conduction in the casting direction is by a factor 10^{-4} smaller than in the other directions, so that we neglect it and start with a model containing 2 space variables and time (which we fix in such a way that the strand enters the secondary cooling zone at t=0). The movement of the strand is modelled in such a way as if D were kept fixed and the casting machine were moved upward with casting speed v, so that the cooling zones move past the cross-section D. Thus, the coordinate system moves with the strand at speed v.

The space coordinates are fixed in such a way that spray cooling takes place in x-direction. For slab casting, where the ratio between the width (in x-direction) and the length (in y-direction) of the cross section D is less than $\frac{1}{3}$, it is justified (within the practically relevant accuracy) to neglect the dependence of the temperature u on y, which further reduces the dimension of our model. Since spray cooling is done in a symmetric way from both sides of the casting machine and the initial temperature is symmetric, the whole temperature field will be symmetric with respect to the center of the strand, so that it suffices to consider only half the strand with a homogenous Neumann boundary condition at the center. Because of the bow in the casting machine, this is a (slight) simplification.

These considerations lead to the following model:

(2.6) $\quad \frac{\partial}{\partial x}(k(u(x,t)) \cdot \frac{\partial u}{\partial x}(x,t)) = \rho \cdot c(u(x,t)) \cdot \frac{\partial u}{\partial t}(x,t)$

$$\text{for } (x,t) \in [0,\tfrac{d}{2}] \times [0,T]$$

(2.7) $\quad k(u(0,t)) \cdot \frac{\partial u}{\partial x}(0,t) = g_v(t) \cdot [u(0,t) - U_w]$

$$\text{for } t \in [0,T_C]$$

(2.8) $\quad k(u(0,t)) \cdot \frac{\partial u}{\partial x}(0,t) = \sigma \cdot \varepsilon \cdot [u(0,t)^4 - U_a^4]$

$$\text{for } t \in [T_C, T_E]$$

(2.9) $\quad \frac{\partial u}{\partial x}(\tfrac{d}{2},t) = 0 \quad \text{for } t \in [0,T]$

(2.10) $u(x,0) = f(x) \quad \text{for } x \in [0,\tfrac{d}{2}]$

(2.11) $u(s(t),t) = u_1 \quad \text{for } t \in [0,T_E]$.

Here, the constants and variables have the following meaning:

$[0,d]$:   cross-section of the strand in x-direction, cooled by spray cooling at $x = 0$

$u(x,t)$:   temperature at time t and position x; see above for the choice of the space coordinate

k: thermal conductivity [W/mK]

ρ: density [kg/m^3]

c: effective heat capacity [Ws/kg K], see (2.5)

g_v: heat transfer function [W/m^{2}K], see below; v: casting speed

U_w: temperature of the water used in spray cooling

T_C, T_E: times when the strand leaves the secondary cooling zone and is completely solid, respectively; these times depend upon v

T: $\max\{T_C, T_E\}$

U_a: temperature of the surrounding air

σ: Stefan-Boltzmann-constant $[W/m^2 K^4]$

ε: emission factor (depends on the material, for steel around 0.8)

u_1: temperature of complete solidification

f: temperature at time $t = 0$; since the strand is already supported by a thin solid layer then, $f(0) < u_1$

$s(t)$: distance of the solidification front from the left boundary at time t; since $f(0) < u_1$, $s(0) > 0$.

The boundary condition (2.7) models the spray cooling, the function g_v (depending on the casting speed v because of our choice of coordinates) contains the information necessary to regulate the spray cooling system. The boundary condition (2.8) models the radiation after the strand leaves the secondary cooling zone. For the (usual) situation that $T_E < T_C$, i. e., that the strand is completely solidified before it leaves the secondary cooling zone, (2.8) is absent. Note that by definition of T_E, $s(T_E) = \frac{d}{2}$.

In contrast to the classical formulation of a Stefan problem, there is no free boundary present in (2.6) - (2.11); (2.6) is valid on a fixed domain across the isotherm described by (2.11). This is computationally advantageous compared to a two-phase Stefan problem. Note that via the enthalpy formulation, also a Stefan problem can be formulated as a problem on a fixed domain (cf. [4,p.217*ff*.], [29]). There, the enthalpy has a jump discontinuity at the temperature of solidification; the relation (2.1) between enthalpy and temperature has to be replaced by an inclusion involving a maximal monotone set-valued function. Incidentally, the Yosida approximation of this maximal monotone function has the same qualitative form as our enthalpy function ([29]).

As mentioned above, the use of a model with only one space variable is justified for slab casting. However, in continuous casting of billets, a two-dimensional model has to be used (c f. [25], where also cooling in the mould and radiation in the secondary cooling zones are taken into account).

In connection with (2.6) - (2.11), we distinguish between two problems:

In the "direct problem", the temparature field u and the solidification front s have to be computed, all other quantities, especially the heat transfer function g_V, being given.

We are interested in the "inverse problem". Given s, we want to compute g_V and u, with all other quantities given. Actually, for the practical problem it suffices to compute a function g_V such that the resulting solidification front is "sufficiently close" to the given one (see Section 3) ("control problem").

While it is our aim to control the solidification front via secondary cooling, other papers report about the control of different quantities in the continuous casting process:

In [29], the spray cooling is used to control the enthalpy. In [25], the temperature and its variation on the surface of the strand are controlled via secondary cooling. In [2], it is suggested to control the surface temperature, the thickness of the solid shell, and the net production rate via secondary cooling; that paper has the form of a project outline, no results are reported yet. We remark that the methodology suggested in [2] is very similar to ours. All these papers (as ours) were motivated by work for the respective national steel industry, which shows that there is currently world-wide interest in these types of problems in steel industry.

Since our problem is an inverse problem, one has the immediate conjecture that it is ill-posed. Recall that due to Hadamard, a problem is called "well-posed" if for all admissible data, a unique solution exists and that solution depends continuously on the data, where the concept of continuity used has to be such that it is meaningful for the actual problem to be solved. Otherwise, a problem is called ill-posed. See [10] for many aspects of inverse and ill-posed problems. It is crucial to find out if one deals with a well-posed or an ill-posed problem before doing numerical work, since discretization of an ill-posed problem without taking the ill-posedness into account usually leads to severe numerical instability. Ill-posed problems have to be treated numerically by "regularization methods"; at least for linear problems, regularization can be achieved by

- changing the spectrum (e. g.: Tikhonov regularization [16])

- projection into a finite-dimensional space, i. e., restricting the number of degrees of freedom (e. g.: least-squares collocation [7], finite elements [24])

- a combination of these two approaches, as always necessary when implementing e. g. Tikhonov regularization numerically (cf. e. g. [11])

- restricting the class of admissible solutions to a compact set.

In order to get a feeling if we are dealing with an ill-posed problem, we look at the following linear problem, which is at least related to our inverse problem:

$$(2.12) \quad \frac{\partial^2 u}{\partial x^2}(x,t) = \frac{\partial u}{\partial t}(x,t) \qquad x \in {]0,1[}, \quad t \in \mathbb{R}$$

$$(2.13) \quad \frac{\partial u}{\partial x}(0,t) = 0 \qquad t \in \mathbb{R}$$

$$(2.14) \quad u(0,t) = G(t) \qquad t \in \mathbb{R}$$

$$(2.15) \quad u(1,t) = F(t) \qquad t \in \mathbb{R} \, .$$

We consider the inverse problem of determining F in (2.15) from (2.12) - (2.14), especially from the given function G. The relation to our problem is the following: x = 0 stands for the center of the strand, where (2.13) holds (cf.(2.9)). If the solidification front is given, the computation of the temperature G at the center of the strand is a well-posed problem, so that any possible ill-posedness must show up in solving (2.12) - (2.15) for F, the surface temperature. Note that (2.15) could be replaced by a Neumann or a mixed boundary condition, which would be closer to (2.7) without changing the essential argument. Thus, (2.12) - (2.15) gives us at least a feeling for the question about the well-posedness of our inverse problem, although many simplifications have been made.

If we denote the Fourier transform with respect to t by $\hat{}$, then we have with G, F as in (2.14) and (2.15) that

$$(2.16) \quad \hat{F}(\omega) = \hat{G}(\omega) \cdot \cosh \sqrt{i\omega}, \qquad \omega \in \mathbb{R}.$$

Now note that as $|\omega| \to \infty$, $|\cosh\sqrt{i\omega}|$ tends to infinity exponentially, which shows together with (2.16) that errors in $\hat{G}$ (and hence in G) with a high frequency are amplified by an arbitrarily high factor, if only the error frequency is high enough; this amplification factor increases very fast with $|\omega|$. This shows that the problem of determining F in (2.12) - (2.15) is severely ill-posed and indicates that the same must at least be conjectured for our inverse problem of determining g_v in (2.6) - (2.11).

The mathematical properties of the full nonlinear problem (2.6) - (2.11) have been studied in [9]. There, only the radiation effect contained in (2.8) has been omitted, since the considerations were restricted to the part of the casting machine where solidification can be influenced, i. e., to the secondary cooling zone.

Under various assumptions (mainly about smoothness and monotonicity) on the functions appearing in (2.6) - (2.11), which seem to be realistic, the following results about (2.6) - (2.11) have been proven in [9], where precise formulations and proofs can be found:

The inverse problem has at most one solution.

The direct problem has exactly one (classical) solution, which also depends continuously on g_v. This continuity is Lipschitz continuity for the temperature field, distances between temperature fields and between heat transfer functions are measured in the uniform norm. If a sequence of heat transfer functions converges uniformly, then the corresponding times of complete solidification converge, and the resulting solidification fronts converge pointwise. Thus, in this sense, the direct problem is well-posed.

If an a-priori bound on $|g_v|$ and on $|\frac{dg_v}{dt}|$ is assumed, then the solution of the inverse problem (as soon as it exists) depends continuously on the data, i. e., on the solidification front, in the following sense: If the times of complete solidification converge and the solidification fronts converge pointwise, then the corresponding heat transfer functions converge uniformly.

Thus, an a-priori bound restores the stability in the inverse problem. This restoration of stability is achieved by restricting the admissible solutions of the inverse problem to a compact set and thus regularizing the problem as mentioned above. This approach does not yield quantitative stability results, i. e., a modulus of continuity.

For a slightly simplified problem, we could prove in [9] that a certain linear functional, namely essentially the integral over the time-interval $[0,T_C]$, of g_v depends on the data in a Lipschitz continuous way. Note that it is frequently the case that the full inverse problem is "more ill-posed" than the problem of finding a specific linear functional of its solution (cf. [1]). However, in our context this specific linear functional is of no practical use.

We are still working on quantitative stability results for our inverse problem. As a first step into this direction, quantitative stability results for the linear inverse problem of finding F in (2.12) - (2.15) have been proven in [12]: Under various a-priori smoothness assumptions on F, logarithmic and Hölder - type moduli of continuity for point-values of F as functions of G (restricted to times prior to this point) have been constructed. Related inverse problems have been studied e. g. in [6], [20], [23].

We close these theoretical considerations by remarking that because of the apparent ill-posedness of our inverse problem, a solution of the "control problem" mentioned above need not be close to a solution of the inverse problem (see [19,p.441] for this aspect for a (linear) inverse Stefan problem). Fortunately, from the practical point of view, we are content with a solution of the control problem, so that our practical problem will not be as severely ill-posed as the theoretical investigation of the inverse problem in [9] and [12] suggests.

3   <u>The Numerical Treatment of the Problem</u>

As can be seen from Fig. 1, the secondary cooling zone in the VOEST-Alpine slab caster consists of 6 zones. In each of these zones, the spray cooling can be regulated independently, but in each of these zones, the water pressure has to be constant. Hence, there are only 6 degrees of freedom, namely the water pressures in the 6 spray cooling zones. Since this fact is likely to regularize the problem (see Section 2), it should be used in a numerical approach. In our case, the engineers originally wanted us to provide g_v "as a function", they wanted to approximate g_v by a suitable step function afterwards. If we we had followed this suggestion, we might have seen severe instabilities (without additional regularization). Thus, we used this "built-in regularization" from the outset.

Let $L_1, \ldots, L_6$ be the lengths of the 6 spray cooling zones. By definition of T_C,

$$(3.1) \quad \sum_{i=1}^{6} L_i = v \cdot T_C$$

holds. Hence, the heat transfer function g_v is defined on the interval $[0, \frac{1}{v} \cdot \sum_{i=1}^{6} L_i]$. Because of the practical situation just described, we assume that g_v has the form

$$(3.2) \quad g_v(t) = g_i \quad \text{for } \sum_{k=1}^{i-1} L_k < v \cdot t \le \sum_{k=1}^{i} L_k, \; i \in \{1, \ldots, 6\}.$$

The 6 "heat transfer coefficients" (and possibly the casting speed v) are now the control variables; they can vary within prescribed ranges:

$$(3.3) \quad b_i \le g_i \le b_{i+7} \qquad (i \in \{1, \ldots, 6\})$$

$$(3.4) \quad b_7 \le v \le b_{14}.$$

The $b_1, \ldots, b_{14}$ are given constants. The restrictions (3.3) –

(3.4) describe a set in $\mathbb{R}^7$, which we denote by G. By C, we denote the function that assigns to each $y := (g_1, \ldots, g_6, v) \in G$ the corresponding heat transfer function g_v defined by (3.2). By S, we denote the solution operator of the direct problem (more precisely, its solidification front part), i. e., $S(g_v)$ is defined as the solidification front s as defined by (2.6) – (2.11) if g_v is used in (2.7). Of course, if g_v is strictly a step function, the existence of $S(g_v)$ cannot be assured. However, since we discretize the problem anyway, we can approximate g_v by a suitably smooth function that is close enough to g_v and coincides with g_v on the discretization points. Because of the well-posedness of the direct problem, different smooth approximations of g_v produce solidification fronts that are as close to each other as one wants as soon as the approximations to g_v are close enough to each other. $S(g_v)$ should be understood in this sense.

By s*, we denote the given solidification front to be approximated by suitably setting the parameters $g_1, \ldots, g_6, v$. Because of (2.10) and (2.11), s* must fulfill $f(s^*(0)) = u_1$. If T_E^* is the time of complete solidification for s*, i. e., the smallest t for which $s^*(t) = \frac{d}{2}$ holds, then a necessary condition for complete solidification within the casting machine is that $b_7 \cdot T_E^*$ does not exceed the total length of the casting machine.

The quality of approximation to s^* is measured by a weighted sum of squares of the pointwise deviations at finitely many (up to 20) points in the casting machine. If we take a constant casting speed, i. e., $b_7 = b_{14}$, which we do from now on, this can be translated into a weighted l^2-norm of the error vector at finitely many times $0 < t_1 < \ldots < t_n = T_E^*$. A solidification front that reaches $\frac{d}{2}$, i. e., complete solidification, before time T_E^*, is continued by $s(t) = \frac{d}{2}$ for $T_E < t \leq T_E^*$ for the purpose of computing the quality of approximation. The given weights reflect the different importance assigned to the accuracy in different parts of the casting machine by the engineers.

Note that since the casting machine is imbedded into the larger environment of steel making and rolling, the casting speed is set from outside requirements and is in practice not available as a control variable, although numerical tests showed its great influence on the solidification front (see [21,pp.192ff.]).

Our control problem is now to determine the heat transfer coefficients such that the quality of approximation (as defined above) to the desired solidification front s* is as good as possible; this leads to the following optimization problem: Given s* , the casting speed v, points $0<t_1<..<t_n=T_E^*$, weights $\gamma_1, \ldots ,\gamma_n>0$ and bounds $b_1, \ldots , b_{14}$ with $b_7=b_{14}$, find $y=(g_1, \ldots , g_6,v)$ fulfilling (3.3), (3.4) such that under these restrictions,

$$(3.5) \quad \sum_{i=1}^{n} \gamma_i [S(C(y))(t_i)-s^*(t_i)]^2 \rightarrow \min.$$

This is a constrained nonlinear optimization problem; the constraints are linear and have the very simple form (3.3) (and (3.4), where in all practical applications we put $b_7=b_{14}$, although the program also allows that $b_7<b_{14}$). Note that each computation of S and hence each evaluation of the objective functional involves solving the direct problem for (2.6) – (2.11) and is hence time consuming. Since we were required by VOEST-Alpine to imbed our program into an existing program environment, we used their finite element program [17] for solving the direct problem, which we sped up for our purposes by removing features not needed for solving (2.6) – (2.11); the solidification front was then computed by interpolation using (2.11).

For solving our linearly constrained optimization problem, we chose Rosen's gradient projection method (with line search) (cf. [28]). The "gradient" of the objective functional was approximated by forward differences; we use quotation marks since we do not know if the objective functional in (3.5) is differentiable, since it involves the operator S. Thus, the question of differentiability of the objective functional involves the question of differentiable dependence of S on g_v in (2.6) – (2.11) (cf. [18] for this question for a linear Stefan problem). Thus, strictly speaking, we work with a gradient-like search direction.

The choice of the line search turned out to be crucial, since the objective function appeared to have quite flat minima, which seems to reflect the underlying ill-posedness of the inverse problem. For details about the line search, the whole algorithm and its performance on test examples see [8,pp.193-194] and [21].

We close with numerical results obtained with our program in actual use with real data in the VOEST-Alpine company. Details and more plots can be found in [21]. The aim was to approximate a fixed solidification front (fixed as a function of position in the casting machine) with different casting speeds. Each example is illustrated by a figure. In each of the figures 3 - 6, the upper half shows the desired solidification front s* as solid line, the initial front of solidification resulting from the starting values of $g_1, \cdots, g_6$ as dotted line, and the resulting solidification front after 3 iterations as broken line. The coordinates in x-direction represent time, the coordinates in y-direction represent the thickness of solid steel and are distorted in such a way that the picture provides maximal information.

The lower half of each figure shows the heat transfer coefficients in the six cooling regions, where the dotted lines represent the starting values and the solid lines represent the results after three iterations; if there is no dotted line, it is hidden under the solid line.

The starting values for $g_1, \cdots, g_6$ have been chosen either as values actually used in practice or by "educated guess" by the engineers.

<u>Example 3.1</u>: see Fig. 3;
casting speed v = 1.5m/min,n = 13; s represents the result after
3 iterations.

t_i [sec]	γ_1	$s(t_i)$ [m]	pointwise relative error compared to s*[%]
34	20	0.0168	0.03
42	20	0.0175	-0.23
60	20	0.0221	0.88
86	10	0.0270	1.19
118	10	0.0319	0.98
170	5	0.0388	2.45
240	5	0.0456	-0.45
340	4	0.0545	-1.31
480	4	0.0650	-1.12
670	3	0.0772	-1.92
950	2	0.0947	0.34
1340	1	0.1154	1.78
1540	0.8	0.1295	0.00

The objective functional in (3.5) was reduced from an initial value of 536 to 23, the resulting heat transfer coefficients were 904, 507, 231, 130, 130, 130 W/Km^2. As can be seen
from Fig. 3, we could not only approximate the desired solidification front to a reasonable accuracy (the maximal relative
error is 2.5 %, i. e., less than 1 mm, which is certainly below the accuracy with which the actual solidification front can
be monitored by one of the methods mentioned in [8], but
also reduce the energy needed for cooling. The point where all
3 curves meet in Fig. 3 is the point where the strand becomes
completely solid.

<u>Example 3.2</u>: see Fig. 4;

casting speed $v = 1.8\text{m/min}$; n, γ_i and s as in Example 3.1.

t_i [sec]	$s(t_i)$ [m]	pointwise relative error compared to s^*[%]
28	0.0164	-2.50
36	0.0171	-2.56
50	0.0215	-1.61
72	0.0264	-1.29
99	0.0311	-1.46
145	0.0369	-2.70
200	0.0452	-1.33
285	0.0547	-0.86
400	0.0656	-0.09
560	0.0779	-1.06
790	0.0944	0.04
1120	0.1140	0.49
1300	0.1295	0.00

The objective function was reduced from an initial value of 439 to 24, the resulting heat transfer coefficients were 1054, 755, 568, 445, 233, 260 W/Km^2. As can be seen in Fig. 4, the result obtained from the initial heat transfer coefficients does not produce complete solidification within the casting machine, so that it could not have been used in practice. Our result allows the use of the relatively high casting speed $v = 1.8\,\text{m/min}$, which is of course of practical relevance.

<u>Example 3.3:</u> see Fig. 5; casting speed $v = 1.7$ m/min, n, γ_i and s as in Example 3.1.

t_i [sec]	$s(t_i)$ [m]	pointwise relative error compared to s*[%]
30	0.0164	-2.51
38	0.0170	-2.58
52	0.0216	-1.59
76	0.0265	-0.80
104	0.0314	-0.72
150	0.0371	-2.23
210	0.0455	-0.62
300	0.0549	-0.55
425	0.0657	0.00
590	0.0781	-0.80
840	0.0944	0.01
1180	0.1121	-1.15
1360	0.1235	-4.67

Here, the objective functional was reduced from an initial value of 113.5 to 27.3, the resulting heat transfer coefficients were 1036, 735, 509, 344, 258, 130 W/Km^2. Compared to Examples 3.1 and 3.2, the reduction in the objective functional is lower, and the highest pointwise error is at the end, which is not desirable . A comparison with Example 3.4 shows the influence of the weights γ_i:

<u>Example 3.4</u>: see Fig. 6; casting speed $v = 1.7$ m/min, n and s as in Example 3.1, all $\gamma_i = 1$.

t_i [sec]	$s(t_i)$ [m]	pointwise relative error compared to s*[%]
30	0.0164	-2.51
38	0.0170	-2.60
52	0.0215	-1.61
76	0.0265	-0.81
104	0.0314	-0.73
150	0.0371	-2.18
210	0.0457	-0.26
300	0.0554	0.34
425	0.0665	1.14
590	0.0801	1.73
840	0.0958	1.48
1180	0.1144	0.85
1360	0.1295	0

The objective functional was reduced from 93.44 to 6.7. Note that due to the different choice of the weights γ_i, this objective functional is different from those in the previous examples. The resulting heat transfer coefficients were 1027, 731 , 507, 383, 294, 130 W/Km^2. A comparison with Example 3.3 clearly shows the influence of the weights γ_i (see also Fig. 5 and 6). Especially, the accuracy at the desired solidification time $t_i = 1360$ is now satisfactory. Note that as in Example 3.2, the result obtained from the initial heat transfer coefficients does not produce complete solidification within the casting machine in Examples 3.3 and 3.4 (see Fig. 5 and 6).

In all examples, the lower and upper bounds in (3.3) were 130 and 1500 W/Km^2. Note that another comparison between Examples 3.3 and 3.4 shows that quite a large change in the heat transfer coefficients is necessary to achieve a comparatively small change in the solidification front, which in our opinion reflects the ill-posedness of the underlying inverse problem.

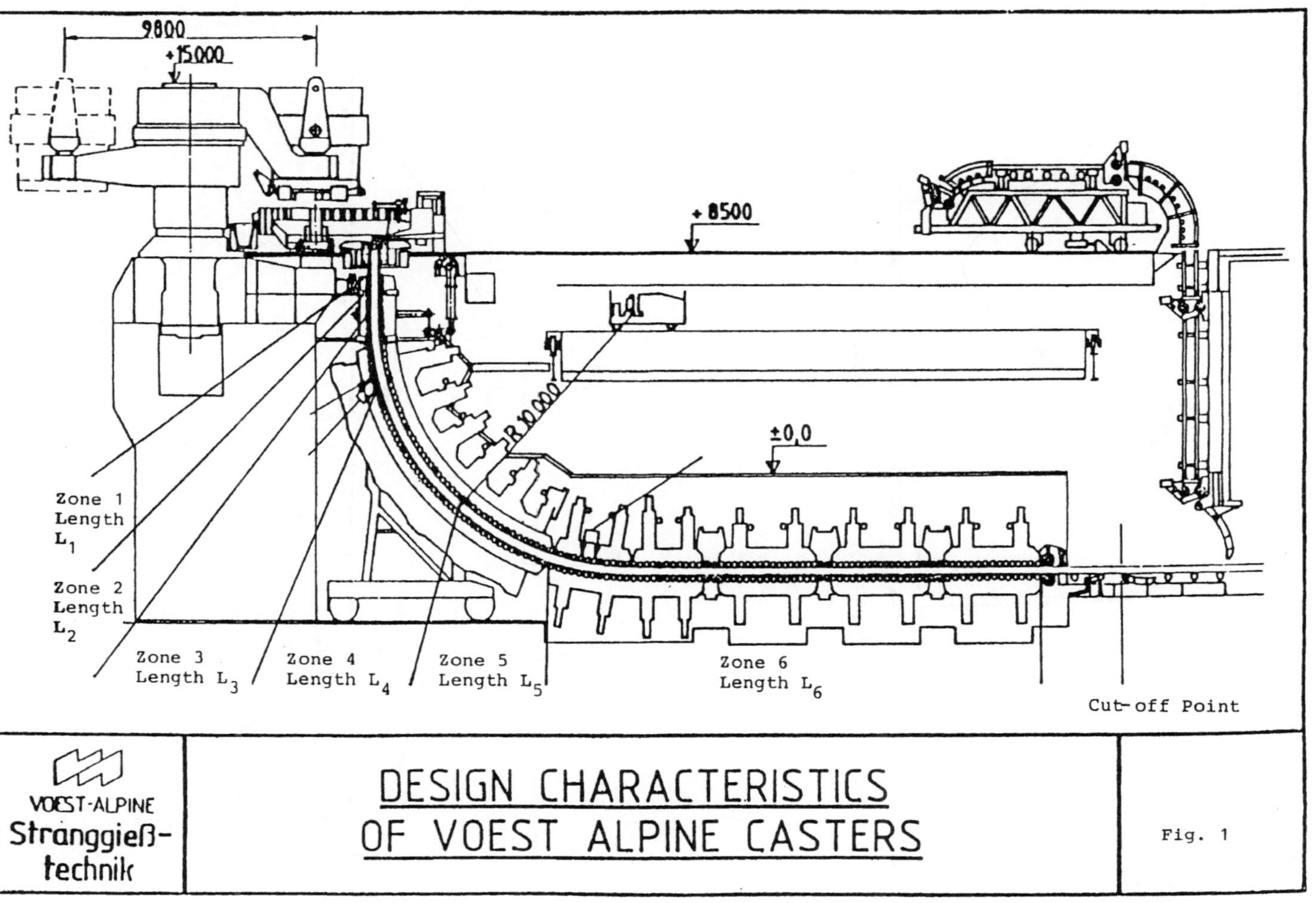

9800
+ 15.000
+ 8500
± 0,0
R 10.000
Zone 1
Length
L_1
Zone 2
Length
L_2
Zone 3
Length L_3
Zone 4
Length L_4
Zone 5
Length L_5
Zone 6
Length L_6
Cut-off Point
VOEST-ALPINE
Stranggieß-technik
DESIGN CHARACTERISTICS
OF VOEST ALPINE CASTERS
Fig. 1

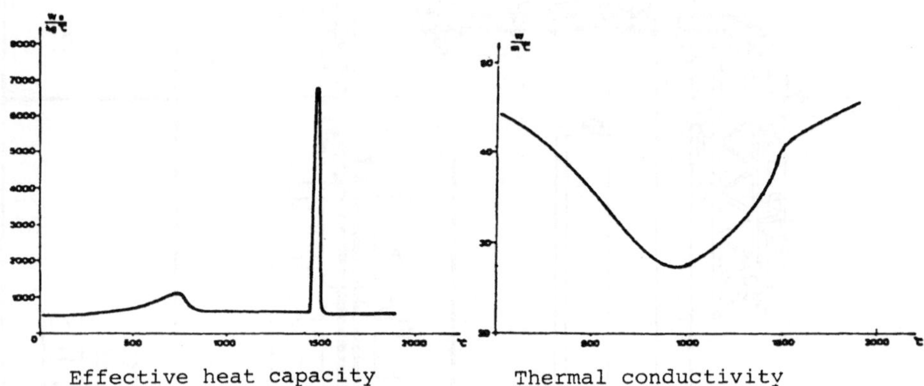

Effective heat capacity

Thermal conductivity

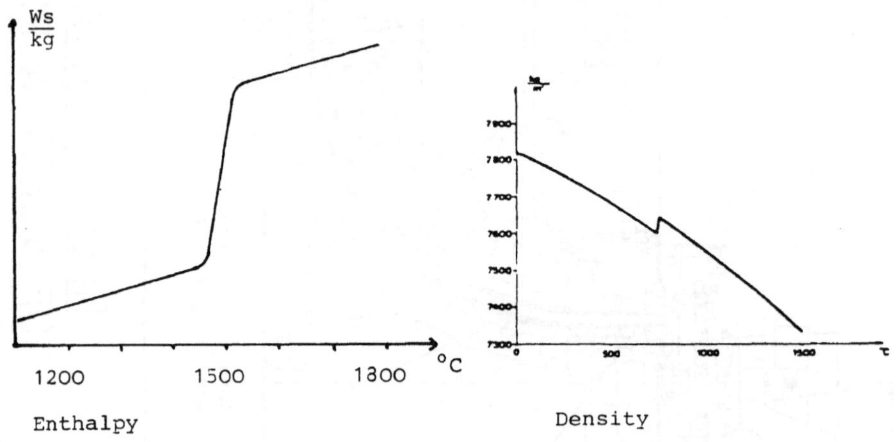

Enthalpy

Density

Fig. 2

60
80
1oo
12o   129.5 = $\frac{d}{2}$
4o
2o
x [mm]
9oo
6oo
3oo
g [W/Km^2]
3oo   6oo   9oo   12oo   15oo
t [sec]

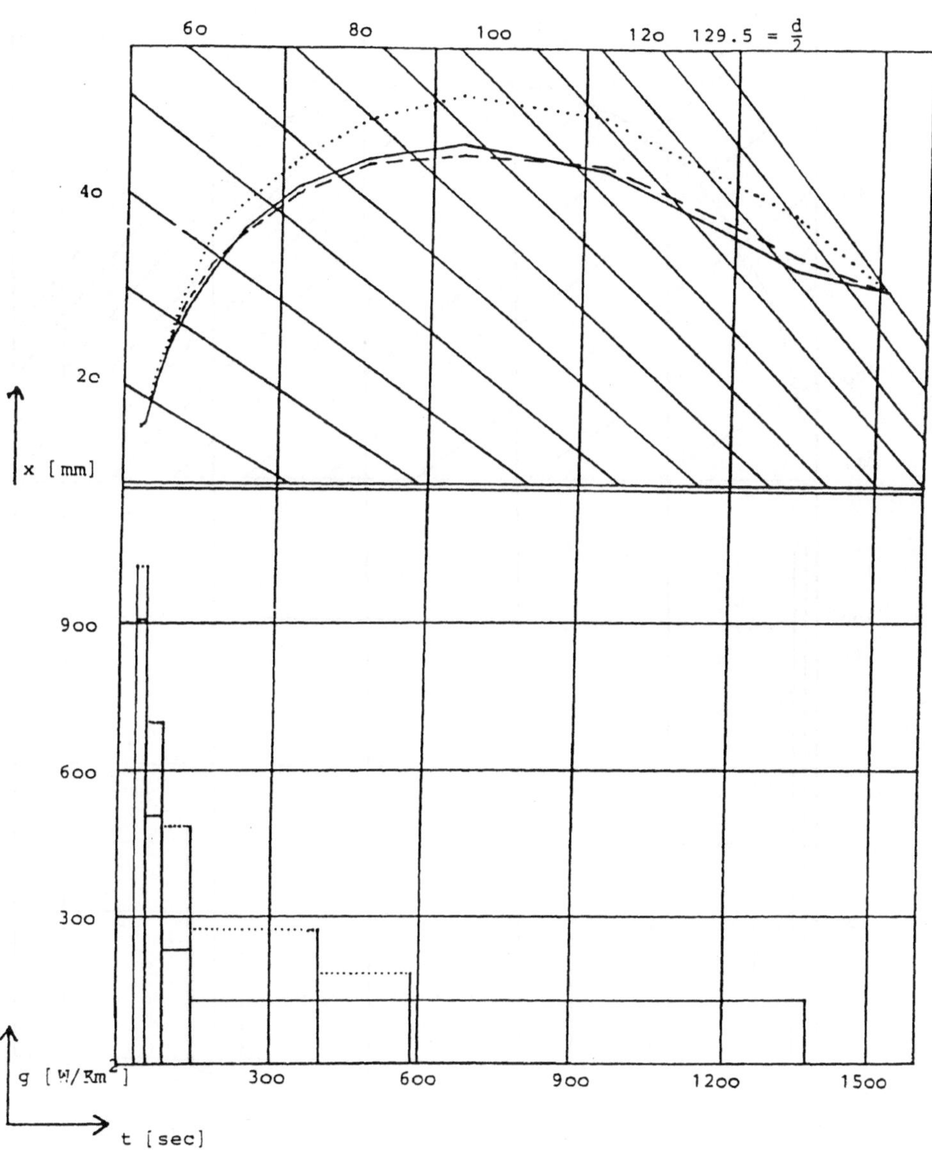

Fig. 3

6o
8o
1oo
12o   129,5 = d/2
4o
2o
x [mm]
9oo
6oo
3oo
g[W/Km²]
3oo
6oo
9oo
12oo
15oo
t [sec]

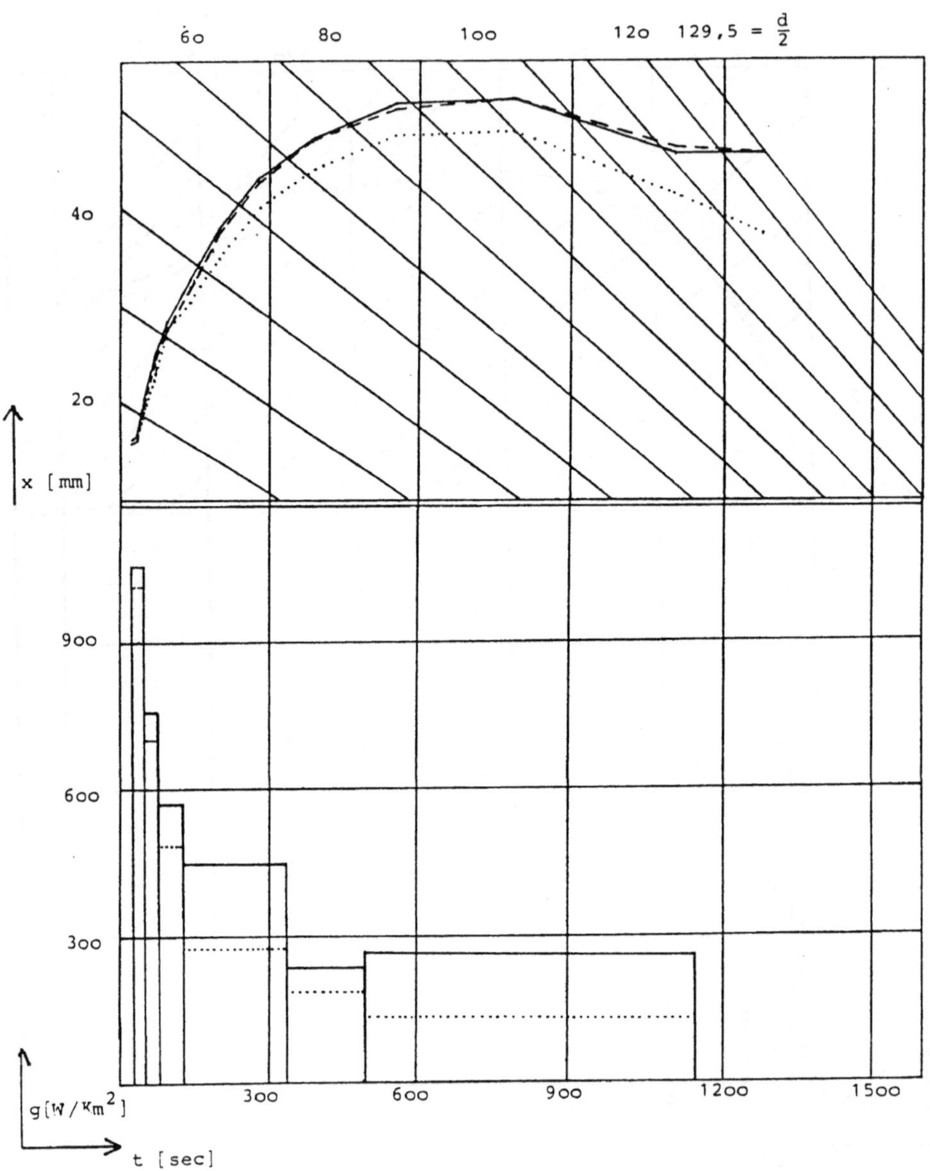

Fig. 4

·60
80
100
120 129·5 = d/2
40
20
x [mm]
900
600
300
g[W/Km²]
300
600
900
1200
1500
t[sec]

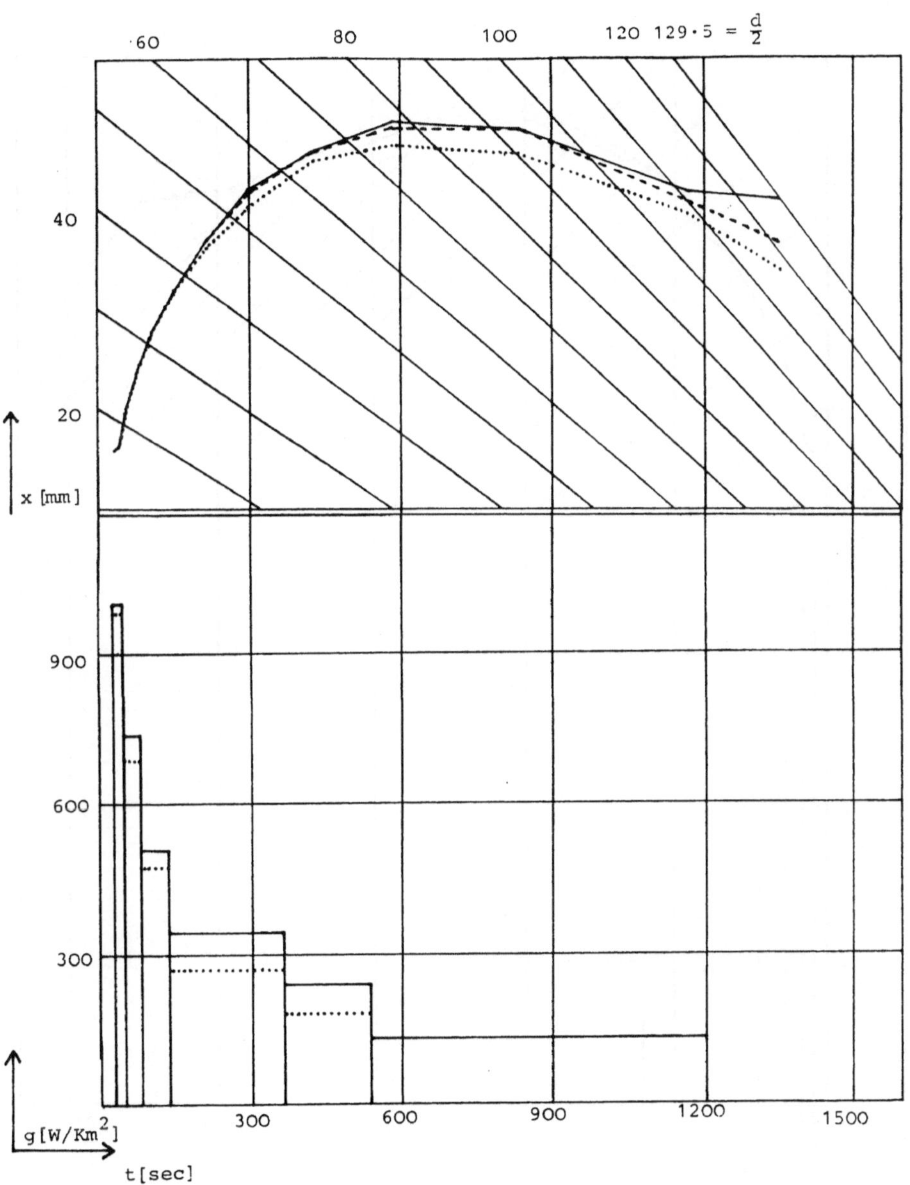

Fig. 5

60
80
100
120 129·5 = $\frac{d}{2}$
40
20
x [mm]
900
600
300
g [W/Km^2]
300
600
900
1200
1500
t [sec]

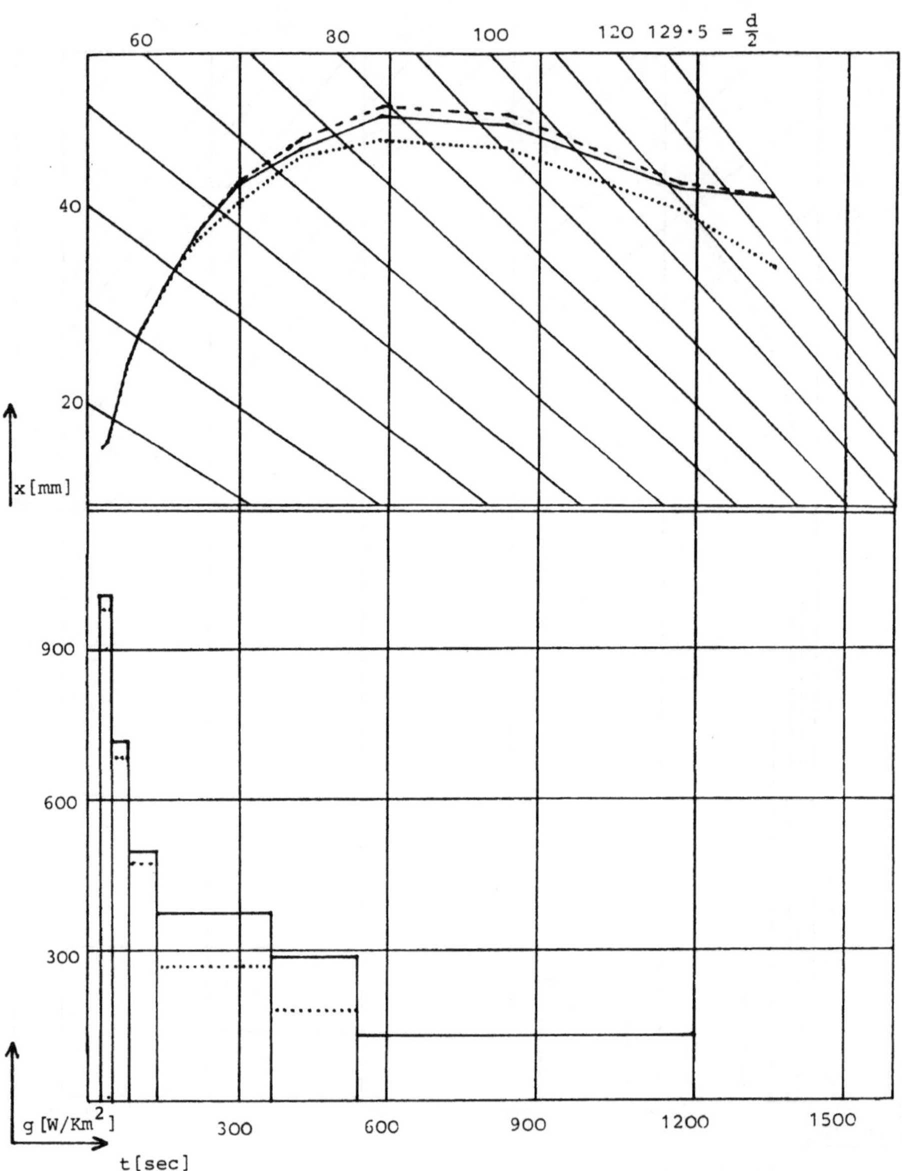

Fig. 6

References:

[1] R. S. Anderssen, The linear functional strategy for im-
     properly posed problems, in : J. Cannon, U. Hornung (eds.),
     Inverse Problems, Birkhäuser, Basel (1986), pp. 11 - 30

[2] N. G. Barton, J. D. Gray (eds.), Proceedings of the 1985
     Mathematics-in-Industry Study Group, CSIRO, Div. of Math.
     and Stat., Australia (1986), pp. 13 - 26

[3] D. Blanchard, M. Fremond, The Stefan problem: Computing
     without the free boundary, Int. Journ. for Num. Meth. in
     Engineering 20 (1984), 757 - 771

[4] J. Crank, Free and Moving Boundary Problems, Clarendon
     Press, Oxford (1984)

[5] A. J. Dalhuijsen, A. Segal, Comparison of finite element
     techniques for solidification problems, Int. Journ. for
     Num. Meth. in Engineering 23 (1986), 1807 - 1829

[6] L. Elden, Approximations for a Cauchy problem for the heat
     equation, Report Li Th-Mat- R -1985-23, Linköping Universi-
     ty (1985)

[7] H. W. Engl, Regularization by least-squares collocation, in:
     P. Deuflhard, E. Hairer (eds.), Numerical Treatment of In-
     verse Problems in Differential and Integral Equations,
     Birkhäuser, Boston (1983), pp. 345 - 354

[8] H. W. Engl, Th. Langthaler, Numerical solution of an in-
     verse problem connected with continuous casting of steel,
     Zeitschrift für Operations Research 29 (1985), B185 - B199

[9] H. W. Engl, Th. Langthaler, P. Manselli, On an inverse pro-
     blem for a nonlinear heat equation connected with conti-
     nuous casting of steel, in: K. H.Hoffmann, W. Krabs (eds.),
     Optimal Control of Partial Differential Equations II,
     Birkhäuser, Basel (1987), 67 - 89

[10] H. W. Engl, C. W. Groetsch (eds.), Inverse and Ill-Posed
     Problems, Academic Press, Boston (1987)

[11] H. W. Engl, A. Neubauer, Convergence rates for Tikhonov
     regularization in finite-dimensional subspaces of Hilbert

scales, Proc. of the Amer. Math. Soc. (to appear)

[12] H. W. Engl, P. Manselli, Stability estimates and regularization for an inverse heat conduction problem in semi-infinite and finite time intervals, in preparation

[13] H. W. Engl, G. Landl, A scheduling problem in the production line "steel making - continuous casting - hot rolling", submitted

[14] A. Fasano, M. Primicerio (eds.), Free Boundary Problems: Theory and Applications, Vol. I and II, Pitman, Boston (1983)

[15] A. Fasano, M. Primicerio, General free boundary problems for the heat equation, J. Math. Anal. Appl. 59 (1977), 694 - 723

[16] C. W. Groetsch, The Theory of Tikhonov Regularization for Fredholm Equations of the First Kind, Pitman, Boston (1984)

[17] C. Jaquemar, Rechnergestützte Auslegung von Aufheiz- und Abkühlvorgängen bei der Stahlerzeugung und Wärmebehandlung, Berg- und Hüttenmännische Monatshefte, Heft 1 (1979), 20 - 30

[18] P. Jochum, Differentiable dependence upon the data in a one-phase Stefan problem, Math. Meth. Appl. Sci. 2 (1980), 73 - 90

[19] P. Knabner, Control of Stefan problems by means of linear-quadratic defect minimization, Numer. Math. 46 (1985), 429 - 442

[20] P. Knabner, S. Vessella, Stability estimates for ill-posed Cauchy problems for parabolic equations, in [10],pp.351 - 368

[21] Th. Langthaler, Numerische Lösung eines inversen Erstarrungsproblems beim Stranggießen von Stahl, Diplomarbeit, Univ. Linz (1985)

[22] B. Lindorfer, K. L. Schwaha, Thermal analysis of the cc process, internal report, VOEST-Alpine, Linz (1984)

[23] P. Manselli, K. Miller, Calculation of the surface temperature and heat flux on one side of a wall form measure-

ments on the opposite side, Ann. Mat. Pura Appl. (IV) 123 (1980), 161 - 183

[24] F. Natterer, The finite element method for ill-posed problems, RAIRO Analyse Num. 11 (1977), 271 - 278

[25] P. Neittaanmäki, On the control of the secondary cooling in the continuous casting process, in: K. H. Hoffmann, W. Krabs (eds.), Optimal Control of Partial Differential Equations II, Birkhäuser, Basel (1987), pp. 161 - 177

[26] I. Pawlow, Y. Shindo, Y. Sakawa, Numerical solution of a multidimensional two-phase Stefan problem, Numer. Funct. Anal. and Optimiz. 8 (1985), 55 - 82

[27] B. Rogberg, Testing and application of a computer program for simulating the solidification process of a continuously cast strand, Scand. Journ. Metallurgy 12 (1983), 13 - 21

[28] J. B. Rosen, The gradient projection method for nonlinear programming (part I: linear constraints), SIAM Journ. Appl. Math. 8 (1960), 181 - 217

[29] C. Saguez, Simulation and optimal control of free-boundary problems, in: J. Albrecht, L. Collatz, K. H. Hoffmann (eds.), Numerical Treatment of Free Boundary Value Problems, Birkhäuser, Basel (1982), pp. 270 - 286

[30] R. E. White, An enthalpy formulation of the Stefan problem, SIAM Journ. Num. Anal. 19 (1982), 1129 - 1157

[31] H. A. Wiesinger, G. N. Holleis, K. L. Schwaha, Concept and design aspects of continuous slab casting machines for hot charging and direct rolling practice, in: Proc. of the 5th Internat. Iron and Steel Congress, Washington, D. C., Iron and Steel Society (1986)

Acknowledgement: This project was supported by the Austrian Fonds zur Förderung der wissenschaftlichen Forschung (project S32/03).

Project 4

OPTIMAL REHEATING OF SLABS IN A PUSHER TYPE REHEATING FURNACE

D.Auzinger, Hj.Wacker

1   Introduction

1.1   Organisation of the Paper

In the article by Engl and Langthaler (this volume) the process of continuous casting of steel is described. If there is no possibility of rolling the slabs immediately after the casting, they must be reheated for the rolling. In the VOEST-ALPINE AG in Linz this is done in pusher type reheating furnaces, which are described in section 1.2.

When the VOEST decided to automate the furnaces in order to guarantee a uniform quality of the rolling products, they consulted us for developing and solving a model for reducing the enormous energy demand. The steady state model for the off-line determination of the optimal reheating strategy was developed together with our partners of the VOEST, mainly Dipl.-Ing.B. Lindorfer. This model and the solution technique are described in chapter 2.

Chapter 3 deals with an efficient method for computing the heat distribution of the slabs and gives some hints for an on-line control. At the moment, such an on-line control based on the results of the off-line optimization is implemented by our partners of the VOEST. When some practical experience will have been gained from the present two step strategy (off-line optimization, on-line control), an instationary on-line optimal control model will be developed.

1.2   <u>Description of the Furnace</u>

Figure 1.1 shows a length cut through the furnace. It is about 36 m long (x-direction), 7 m high (y-direction) and 13 m wide (z-direction). It consists of 4 zones, each being divided into an upper and a lower part. The slabs, which are blocks of steel of about 1,2 m × 0,2 m × 12 m, are pushed through the furnace with their length-direction in the z-direction of the furnace. The furnace is filled with about 3o slabs, which form a "slab band". Whenever a slab is pushed into the furnace, the slab band moves in x-direction and another slab leaves the furnace at the discharging end, at a temperature of about 125o$^{\circ}$C.

In the first three zones (convective zone, preheating zone, heating zone) the slabs are pushed on water cooled skids, which cause the "skid marks" (compare chapter 3, Fig. 3.3). To diminish these skid marks, the lower part of zone 4 (soaking zone) is an unheated soaking hearth. The furnace is heated by coke gas and by natural gas. There are huge burners at the upper and the lower part of zones 2 and 3, which are the main heating zones, and there are a lot of small burners at the upper part of zone 4. The flue gas streams countercurrently to the slabs. In zone 1 there are no burners, but it is heated by the flue gas coming from zone 2. The gas leaves the furnace at the charging point and is used to preheat the air required for the combustion in a recuperator.

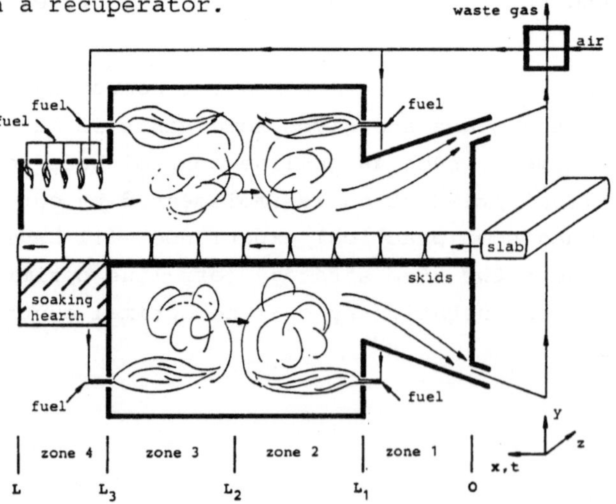

Fig. 1.1
pusher type
reheating furnace

2   The Optimal Reheating Process

2.1   The Reheating Model

Our reheating model involves three parts: the heat equation for the slabs, the flue gas model and the model for the recuperator. For the optimization model we obtain a fourth part consisting of the objective and technical restrictions. For ease of presentation the model is somewhat simplified.

2.1.1 Basic Assumptions for the Model. Some assumptions are required to idealize and simplify the real reheating process, which is much too complex to be modelled "exactly". The first assumption is the strongest:

(A1)   steady operation of the furnace: slabs of the same size and quality are charged at a constant temperature. They are discharged after a constant heating time and at a constant temperature.

(A2)   the real movement of the slabs is replaced by a movement with constant velocity v.

(A3)   the length direction of the slabs (z-direction) is neglected.

(A4)   within the slabs no heat conduction in x-direction is considered, as this conductive heat transport is small compared to the other heat transport effects.

(A5)   the flue gas in the zones 2 and 3 is "well-stirred", which means that the gas of the whole zone has a uniform temperature.

(A6)   for zones 1 and 4 we assume a differential flue gas model, neglecting the y- and the z-direction. The reason for the different types of flue gas models is the furnace geometry and the position of the burners.

2.1.2 The Heat Equation for the Slabs. The real reheating process takes place in a four-dimensional space (x,y,z,t), but our assumptions (A1), (A2), (A3) enable us to define our model in a two dimensional (x,y)-space.

The resulting heat equation for the slabs is

(slab: h.e.)

$$\frac{\partial}{\partial x}\left(\rho.v.c_m(T)T\right) = \frac{\partial}{\partial y}\left[\lambda(T)\frac{\partial T}{\partial y}\right].$$

$T(x,y)$ for $x \in [0,L] \times [-d,d]$ is the heat distribution of the slabs, ρ is the density and $c_m(T)$ the average specific heat of steel, v the velocity of the slabs and $\lambda(T)$ the conductivity of steel. L is the length of the furnace, d half the height of the slabs. The initial value for T is

(slab: i.v.)

$$T(0,y) = T_A \qquad\qquad y \in [-d,d],$$

the boundary conditions depend on the position of the slab. For the upper bound of the slabs ($y = d$) we obtain

(slab: u.b.)

$$\lambda(T)\frac{\partial T}{\partial y} = \begin{cases} \alpha_1^u(\theta_1^u-T)+c_1^u(\theta_1^{u\,4}-T^4) & x\in(0,L_1] \\[2ex] \alpha_2^u(\theta_2^u-T)+c_2^u(x)(\theta_2^{u\,4}-T^4)+c_3^u(x)(\theta_3^{u\,4}-T^4) & x\in(L_1,L_2] \\[2ex] \alpha_3^u(\theta_3^u-T)+c_2^u(x)(\theta_2^{u\,4}-T^4)+c_3^u(x)(\theta_3^{u\,4}-T^4) & x\in(L_2,L_3] \\[2ex] \alpha_4^u(\theta_4^u-T)+c_4^u(\theta_4^{u\,4}-T^4) & x\in(L_3,L] \end{cases}$$

The left hand side denotes the heat transport from the surface into the interior, the right hand side is the heat received by the surface. In zone 1 we have a convective term with α_1^u the convectivity constant and $\theta_1^u(x)$ the upper flue gas temperature, and a radiative term with radiation constant c_1^u. In zone 2 we have a convective and a radiative heat transfer from the flue gas of zone 2 (well stirred, temperature θ_2^u) to the slabs. Due to the furnace geometry we also have to consider a radiative heat exchange with the flue gas of zone 3 ("cross radiation"). Zone 3 is symmetrical to zone 2, and for zone 4 we get analogous boundary conditions as for zone 1.

For the _lower_ _bound_ of the slabs we get similar conditions (slab: l.b.), the heat losses of the slabs caused by the water cooled skids are considered by an additional term. At zone 4 we have no heat exchange with a flue gas, but heat losses through the soaking hearth.

2.1.3 <u>The Flue Gas Model.</u> For given flue gas temperature the slab model presented in 2.1.2 describes the full reheating process. This fact is used in chapter 3 where we discuss an on-line model based on measured furnace temperatures. For the optimization we have to model the flue gas. To avoid too much complexity - Navier Stokes equation with combustion, for instance - we use an one-dimensional description of the flue gas temperature: $\theta(x)$. We balance both mass and energy.

As the flue gas moves countercurrently to the slabs, we start at the discharging end (zone _4_, _upper_ part) and get for the _mass_ _stream_:

(gas 4u: m.s.)

$$mg_4^u(x) = [B_{EX} + \frac{L-x}{L-L_3} B_4^u] (1+\nu_4)$$

Here we assume that the total fuel input B_4^u is distributed over the whole length of the zone, the resulting "fuel input density" is $\frac{B_4^u}{L-L_3}$. B_{EX} is the given fuel input at the discharging point, which is to make up for the heat losses when a slab is discharged. The term $(1+\nu_4)$ says that for 1 kg fuel ν_4 kg air are required for the combustion.

The equation for the flue gas temperature is

(gas 4u: g.t.)

$$- \frac{d}{dx} (mg_4^u(x) c_m^G(\theta_4^u(x)) \theta_4^u(x)) =$$

$$= \frac{B_4^u}{L-L_3} [H_4 + \nu_4 c_m^A(\theta^A) \theta^A] - 1\alpha_4^u(\theta_4^u - T) - 1c_4^u(\theta_4^{u^4} - T^4) - k_4^W(\theta_4^u - T^H)$$

with the initial value

(gas 4u: i.v.)

$$\theta_4^u(L) = \bar{\theta}_4^u$$

The left hand side of equation (gas 4u: g.t.) is the variation of the heat of the flue gas stream. The right hand side balances the heat streams to and from the flue gas: the first term denotes the heat input by the combustion (H_4 is the heating value of the fuel) and by the input of the preheated air (temperature θ^A). The second and the third term stand for the heat going into the slabs by convection and radiation (1 is the length of the slabs, e.g. 1 = 12 m). The last term denotes the heat losses through the furnace wall.

Zone 3 is assumed to be well stirred, and thus we have an overall heat balance:

(gas 3u: h.b.)

$$mg_4^u(L_3) c_m^G(\theta_4^u(L_3)) \theta_4^u(L_3) + B_3^u(H_3 + \nu_3 c_m^A(\theta^A) \theta^A) =$$

$$= 1 \int_{L_2}^{L_3} [\alpha_3^u(\theta_3^u - T) + c_3^u(x)(\theta_3^{u^4} - T^4)] dx + 1\int_{L_1}^{L_2} c_3^u(x)(\theta_3^{u^4} - T^4) dx$$

$$+ k_3^W(\theta_3^u - T^H) + mg_3^u c_m^G(\theta_3^u) \theta_3^u.$$

The left hand side is the heat entering the zone: flue gas coming from zone 4, combustion and input of air. The right hand side is the heat leaving the zone. The first two terms denote the heat going into the slabs of zone 3 and zone 2 (cross radiation!), the third gives the heat losses through the furnace wall and the last term stands for the heat leaving with the flue gas. This flue gas mass stream is

(gas 3u: m.s.)

$$mg_3^u = mg_4^u(L_3) + B_3^u(1+\nu_3).$$

For the upper part of zone 2 we get a heat balance equation (gas 2u: h.b.) analogous to (gas 3u: h.b.), the mass stream leaving zone 2 is mg_2^u.

The equation for the gas temperature of zone 1 is similar to that of zone 4, but without fuel input ($B_1^u = 0$). Thus we have the constant flue gas mass stream mg_2^u. The initial value for equation (gas 1u: g.t.) is

(gas 1u: i.v.)

$$\theta_1^u(L_1) = \theta_2^u.$$

At the lower part of zone 4 we have the soaking hearth and thus no flue gas. For the lower parts of the first three zones we have similar models as for the upper parts, they are described by the sets of equations (gas 3l: h.b., m.s.), (gas 2l: h.b., m.s.) and (gas 1l: g.t., i.v.).

2.1.4 The Recuperator Model. The flue gas leaving the furnace at the charging point is used to preheat the air required for the combustion. This is done in a recuperator (heat exchanger) and helps to save energy. Before entering the recuperator, the upper and the lower gas stream are mixed, the resulting flue gas temperature is θ^G:

(recu: 1)

$$mg_2^u c_m^G(\theta_1^u(0)) \; \theta_1^u(0) + mg_2^l c_m^G(\theta_1^l(0)) \theta_1^l(0) =$$

$$= (mg_2^u + mg_2^l) \, c_m^G(\theta^G) \theta^G.$$

Within the recuperator there is a counterstream of flue gas, which is cooled from temperature θ^G to θ^{WG}, and air, which enters at a given temperature $\bar{\theta}_A$ and leaves with the higher temperature θ_A. The heat exchanger is described by a boundary value problem for ordinary differential equations, which can be solved explicitely. The exit temperatures θ^{WG} and θ^A satisfy the two equations

(recu: 2,3)

$$(mg_2^u + mg_2^l) \, [c_m^G(\theta^G) \theta^G - c_m^G(\theta^{WG}) \theta^{WG}] =$$

$$= m^A [c_m^A(\theta^A) \theta^A - c_m^A(\bar{\theta}^A) \bar{\theta}^A] =$$

$$= k^R \frac{(\theta^G - \theta^A) - (\theta^{WG} - \bar{\theta}^A)}{\ln(\theta^G - \theta^A) - \ln(\theta^{WG} - \bar{\theta}^A)}$$

The first line denotes the heat leaving the gas stream, the second line the heat received by the air and the last line stands for the heat being transferred within the recuperator. m^A is the quantity of air required, it is a linear function of the fuel streams B: $m^A = \nu_2(B_2^u + B_2^l) + \nu_3(B_3^u + B_3^l) + \nu_4(B_4^u + B_{EX})$.

2.1.5 <u>Discussion of the Reheating Model.</u> The set of equations (slab), (gas) and (recu) describes the reheating process for given fuel streams B. It is a coupled system of a partial differential equation, some ordinary differential equations and some nonlinear algebraic equations. Figure 2.1 shows the way they are coupled: All through the furnace there is a counter-

stream of slabs and flue gas, which influence one another. There
is an initial condition for the slabs at the charging point and
an initial condition for the flue gas at the discharging point.
In addition, the gas temperature at the charging point influences the flue gas of the whole furnace via the preheated air.
Thus, the three parts of the model cannot be decoupled and solved
separately, the system is a boundary value problem (BVP) connected
with some nonlinear equations.

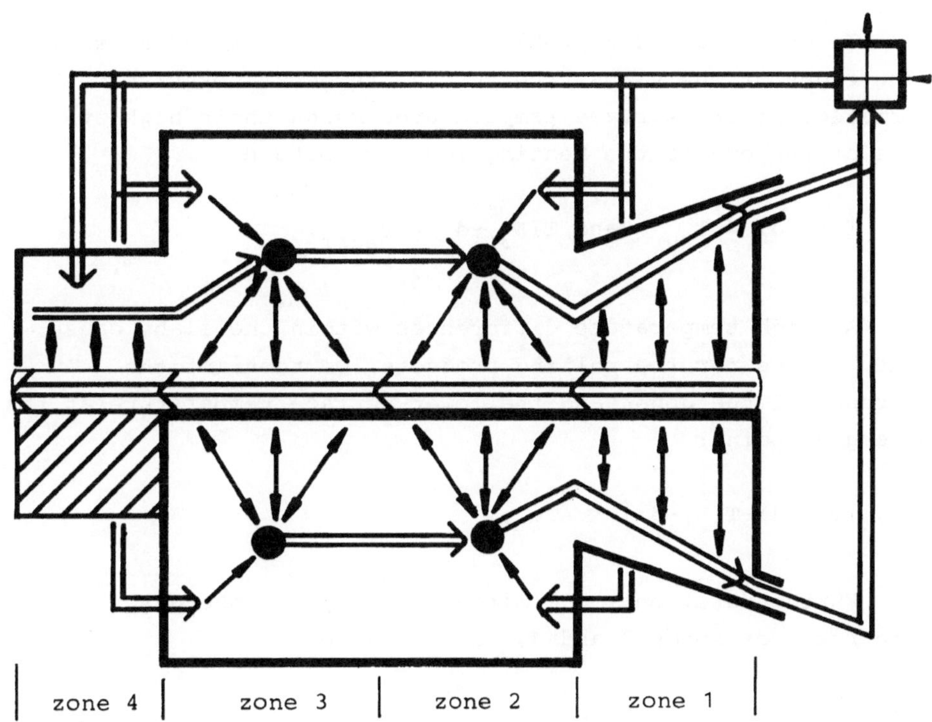

Fig. 2.1   system slabs - flue gas - recuperator

2.2   The Optimization Model

2.2.1 Additional Restrictions for the Reheating Process.

An optimal reheating strategy has to satisfy some additional restrictions. The most important one is a condition for the discharging temperature:

$$(2.1)\quad \int_{-d}^{d} c_m(T(L,y))T(L,y)\,dy = 2d\, c_m(T_{discharge})T_{discharge}$$

In order to avoid problems with the tinder and a melting of the edges of the slabs, the surface temperatures have to be restricted. As the surface temperatures reach their highest values at the end of the heating zone, we obtain

$$(2.2)\quad T(L_3,d) \leq T_{surf} \text{ and } T(L_3,-d) \leq T_{surf}.$$

As high temperature differences within the slabs cause quality losses for the rolling products, we restrict the difference of the upper and the lower surface temperature at the discharging point:

$$(2.3)\quad |T(L,d)-T(L,-d)| \leq \Delta T_{surf}$$

Furthermore, only a limited quantity of coke gas, which is the fuel for zones 3 and 4, is available:

$$(2.4)\quad B_3^u + B_3^l + B_4^u + B_{EX} \leq B_{coke}$$

One also has restrictions for the fuel streams because of the constrained burner capacity and the fact that the flames must not get extinguished:

$$(2.5)\quad \underline{B}_2^u \leq B_2^u \leq \bar{B}_2^u, \text{ analogously for } B_2^l,\ B_3^u,\ B_3^l,\ B_4^u.$$

The objective is the energy demand resp. the costs, which is to be minimized:

$$(2.6) \quad a_2(B_2^u + B_2^l) + a_3(B_3^u + B_3^l) + a_4 B_4^u = \text{Min!}$$

2.2.2 Discretization of the Heat Equation. The model for the optimal reheating presented in chapters 2.1 and 2.2.1 will now be transformed into the following standard optimization problem

$$(2.7) \quad \begin{aligned} f(x) &= \text{Min!} \\ g_j(x) &= 0 \quad \text{for} \quad j = 1,\ldots,m1 \\ g_j(x) &\geq 0 \quad \text{for} \quad j = m1+1,\ldots,m \end{aligned}$$

As a first step we discretize the heat equation. Exploiting the relation

$$c_m(T) = \frac{1}{T} \int_0^T c_p(\tau)\,d\tau$$

between the average specific heat c_m and the specific heat c_p, the heat equation (slab: h.e.) can be made explicit:

$$(2.8) \quad \frac{\partial T}{\partial x} = \frac{1}{\rho v c_p(T)} \frac{\partial}{\partial y}\left(\lambda(T)\frac{\partial T}{\partial y}\right)$$

For the discretization of the term $\frac{\partial}{\partial y}\left(\lambda(T)\frac{\partial T}{\partial y}\right) = \frac{\partial}{\partial y}\lambda(T)\frac{\partial T}{\partial y} + \lambda(T)\frac{\partial^2}{\partial y^2}T$ we consider the points $y_1 := d,\ldots y_i := y_{i-1} - h,\ldots$ $y_M := -d$ with $h = \frac{2d}{M-1}$ and $T_i(x) \sim T(x,y_i)$ for $i = 1,\ldots,M$. Writing

$$(2.9) \quad \lambda(T)\frac{\partial T}{\partial y} = r^u(x,T) \quad \text{and} \quad -\lambda(T)\frac{\partial T}{\partial y} = r^l(x,T)$$

for the boundary conditions (slab: u.b., l.b.), and using a centraldifferential quotient for both the first and the second derivative, the heat equation becomes

$$(2.1o) \quad \frac{d}{dx} T_1(x) = \frac{1}{4h^2 \rho v c_p(T_1)} \{ [\lambda(T_2 + \frac{2h}{\lambda(T_1)} r^u(x,T_1)) - \lambda(T_2)] \cdot$$

$$\cdot [\frac{2h}{\lambda(T_1)} r^u(x,T_1)] + 4\lambda(T_1)[2T_2 + \frac{2h}{\lambda(T_1)} r^u(x,T_1) - 2T_1] \}$$

$$(2.11) \quad \frac{d}{dx} T_i(x) = \frac{1}{4h^2 \rho v c_p(T_i)} \{ [\lambda(T_{i+1}) - \lambda(T_{i-1})][T_{i+1} - T_{i-1}] +$$

$$+ 4\lambda(T_i)[T_{i+1} - 2T_i + T_{i-1}] \} \qquad \text{for } i = 2,\ldots,M-1$$

$$(2.12) \quad \frac{d}{dx} T_M(x) = \frac{1}{4h^2 \rho v c_p(T_M)} \{ [\lambda(T_{M-1} + \frac{2h}{\lambda(T_M)} r^l(x,T_M)) - \lambda(T_{M-1})] \cdot$$

$$\cdot [\frac{2h}{\lambda(T_M)} r^l(x,T_M)] + 4\lambda(T_M)[2T_{M-1} + \frac{2h}{\lambda(T_M)} r^l(x,T_M) - 2T_M] \}$$

The tridiagonal system (2.1o), (2.11), (2.12) is an order two approximation of the heat equation.

The differential equations (gas 4u: g.t.), (gas 1u: g.t.) and (gas 1l: g.t.) are made explicit, too. The resulting ODE for zone 4 is

$$(2.13) \quad \frac{d}{dx} \theta_4^u(x) = \frac{1}{mg_4^u c_p^G(\theta_4^u)} \{ l\alpha_4^u(\theta_4^u - T_1) + lc_4^u(\theta_4^{u\,4} - T_1^4) +$$

$$+ k_4^W(\theta_4^u - T^H) + \frac{B_4^u}{L - L_3} [(1 + \nu_4) c_m^G(\theta_4^u)\theta_4^u - H_4 - \nu_4 c_m^A(\theta^A)\theta^A] \}.$$

Equations (2.14) and (2.15) for the upper and the lower flue gas temperature at zone 1 are similar.

2.2.3 <u>Transformation into a Standard Optimization Problem.</u> The reheating model resulting from chapter 2.2.1 is dominated by an ODE-BVP for the temperatures $T_i(x)$, $i = 1,\ldots,M$ and the flue gas temperatures $\theta(x)$. Both the charging temperature of the slabs and the gas temperature at the discharging point are known. The heat balance equations for zones 2 and 3 can easily be written as ODE's and can be incorporated into the system.

To solve the problem we introduce the additional unknowns $\theta_o^u := \theta_1^u(0)$ and $\theta_o^l := \theta_1^l(0)$ for the temperatures of the flue gas streams leaving the furnace. These unknowns and the charging temperatures define a full set of initial values for the system (slab), (gas) at zone 1. The unknown values θ_o^u and θ_o^l are implicitly defined by the two conditions (gas 1u: i.v.) and (gas 1l: i.v.). For reasons of stability we do not use the end temperatures of zone 1 for the initial values of zone 2, but we introduce additional unknowns $(T_{2,i})_{i=1,\ldots,M}$ and require the continuity conditions $T_{2,i} = T_i(L_1)$ (compare with the multiple shooting method [7]). This allows us a stable integration of the heat equation at zone 2.

Zone 3 is treated completely analogously, the initial values introduced are $(T_{3,i})_{i=1,\ldots,M}$. For zone 4 we also define initial values for the slab temperatures, but for reasons of stability it is not possible to integrate the gas temperature equation against the gas mass stream, as it is done in zone 1. Thus zone 4 is a BVP itself, which has to be solved by suitable methods.

Summing up we obtain an iterative process for solving the reheating model for given B:

- Choose 3M+2 suitable initial values $T_{2,i}$, $T_{3,i}$, $T_{4,i}$, θ_o^u, θ_o^l

- Solve the three IVP's for zones 1,2,3 and the BVP for zone 4

- Improve the estimates of the initial values until the 3M+2 continuity conditions are satisfied.

To avoid a two stage iteration - the optimization itself is an iterative process, too - the continuity conditions resp. the unknown initial values are considered as equality restrictions resp. unknowns of the optimization problem. Of course, this increases the dimension of the optimization problem, but it saves evaluations of the ODE's, which are the main computational amount for the problem.

The described transformation of the model was chosen after a numerical stability analysis of the problem and in order to minimize the computational amount for the evaluation of the system.

2.2.4 The Optimization Problem. The $3M+14$ unknowns of the problem are

B_2^u, B_2^l, B_3^u, B_3^l, B_4^u	fuel streams
θ^A	temperature of preheated air
θ^G, θ^{WG}	gas temperature before and after recuperator
θ_o^u, θ_o^l	gas temperature at charging point
θ_2^u, θ_2^l, θ_3^u, θ_3^l	gas temperature at zones 2 and 3
$T_{2,i}$, $T_{3,i}$, $T_{4,i}$	slab temperatures at the beginning of zones 2,3 and 4

The gas and the slab temperatures $\theta_1^u(x)$, $T_i(x)$, $\theta_1^l(x)$ are solutions of the IVP (2.14), (2.1o), (2.11), (2.12), (2.15) with the initial values θ_o^u, T_{charge}, θ_o^l. They depend on the unknowns θ_o^u, θ_o^l and on the fuel streams B (the gas mass streams are functions of the fuel streams). The slab temperatures of zones 2 resp. 3 are solutions of the IVP (2.1o), (2.11), (2.12) with initial values $T_{2,i}$ resp. $T_{3,i}$. Beside the initial values they depend on the zone temperatures θ_2^u, θ_2^l, θ_3^u, θ_3^l. The gas and slab temperatures at zone 4 are obtained by solving the BVP (2.13), (2.1o), (2.11), (2.12) with $T_i(L_3) = T_{4,i}$ and $\theta_4^u(L) = \bar{\theta}_4^u$. They depend on the unknowns $T_{4,i}$, B_4^u and θ^A.

<u>Constraints of the model:</u> Conditions (gas 1u: i.v.) and (gas 1l: i,v.) give

(eg.1) $\theta_1^u(L_1) - \theta_2^u = 0$

(eg.2) $\theta_1^l(L_1) - \theta_2^l = 0$

The heat balances for zones 2 and 3 - (gas 2 u,l: h.b.) and (gas 3 u,l: h.b.) - become

$$\text{(eg.3)} \quad mg_3^u c_m^G(\theta_3^u)\theta_3^u + B_2^u(H_2 + \nu_2 c_m^A(\theta^A)\theta^A) - 1\int_{L_1}^{L_2} \alpha_2^u(\theta_2^u - T_1) +$$

$$+ c_2^u(x)(\theta_2^{u^4} - T_1^4)dx - 1\int_{L_2}^{L_3} c_2^u(x)(\theta_2^{u^4} - T_1^4)dx - k_2^W(\theta_2^u - T^H) -$$

$$- mg_2^u c_m^G(\theta_2^u)\theta_2^u = 0$$

$$\text{(eg.4)} \quad mg_3^l c_m^G(\theta_3^l)\theta_3^l + B_2^l(H_2 + \nu_2 c_m^A(\theta^A)\theta^A) - 1\int_{L_1}^{L_2} \alpha_2^l(\theta_2^l - T_M) +$$

$$+ c_2^l(x)(\theta_2^{l^4} - T_M^4)dx - 1\int_{L_2}^{L_3} c_2^l(x)(\theta_2^{l^4} - T_M^4)dx - k_2^W(\theta_2^l - T^H) -$$

$$- mg_2^l c_m^G(\theta_2^l)\theta_2^l = 0$$

$$\text{(eg.5)} \quad mg_4^u(L_3) c_m^G(\theta_4^u(L_3))\theta_4^u(L_3) + B_3^u(H_3 + \nu_3 c_m^A(\theta^A)\theta^A) -$$

$$- 1\int_{L_1}^{L_2} c_3^u(x)(\theta_3^{u^4} - T_1^4)dx - 1\int_{L_2}^{L_3} \alpha_3^u(\theta_3^u - T_1) + c_3^u(x)(\theta_3^{u^4} - T_1^4)dx -$$

$$- k_3^W(\theta_3^u - T^H) - mg_3^u c_m^G(\theta_3^u)\theta_3^u = 0$$

$$(eg.6) \quad B_3^1(H_3 + \nu_3 c_m^A(\theta^A)\theta^A) \; - \; 1 \int_{L_1}^{L_2} c_3^1(x)(\theta_3^{1^4} - T_M^4)\,dx \; -$$

$$-1 \int_{L_2}^{L_3} \alpha_3^1(\theta_3^1 - T_M) \; + \; c_3^1(x)(\theta_3^{1^4} - T_M^4)\,dx \; - \; k_3^W(\theta_3^1 - T^H) \; -$$

$$-mg_3^1 c_m^G(\theta_3^1)\theta_3^1 \; = \; 0$$

Recuperator: (recu: 1,2,3)

$$(eg.7) \quad mg_2^u c_m^G(\theta_o^u)\theta_o^u - mg_2^1 c_m^G(\theta_o^1)\theta_o^1 - (mg_2^u + mg_2^1) c_m^G(\theta^G)\theta^G \; = \; 0$$

$$(eg.8) \quad (mg_2^u + mg_2^1)[c_m^G(\theta^G)\theta^G - c_m^G(\theta^{WG})\theta^{WG}] \; -$$

$$-m^A[c_m^A(\theta^A)\theta^A \; - \; c_m^A(\bar{\theta}^A)\bar{\theta}^A] \; = \; 0$$

$$(eg.9) \quad m^A[c_m^A(\theta^A)\theta^A \; - \; c_m^A(\bar{\theta}_A^A)\bar{\theta}_A^A] \; -$$

$$-k^R \frac{(\theta^G - \theta^A) - (\theta^{WG} - \bar{\theta}^A)}{\ln(\theta^G - \theta^A) - \ln(\theta^{WG} - \bar{\theta}^A)} \; = \; 0$$

The discretized condition for the discharging temperature - (2.1) - is:

$$(eg.10) \quad \frac{1}{M-1}\left[\frac{1}{2} c_m(T_1)T_1 \; + \; \sum_{i=2}^{M-1} c_m(T_i)T_i \; + \; \frac{1}{2} c_m(T_M)T_M\right] \; -$$

$$-c_m(T_{discharge})T_{discharge} \; = \; 0$$

Continuity conditions for the slabs:

$$(eg.i+10)_{i=1,\ldots,M} \quad T_i(L_1) - T_{2,i} \; = \; 0$$

$$(eg.\ i+M+1o)_{i=1,\ldots,M} \qquad T_i(L_2)-T_{3,i} = 0$$

$$(eg.\ i+2M+1o)_{i=1,\ldots,M} \qquad T_i(L_3)-T_{4,i} = 0$$

The inequality constraints (2.2), (2.3), (2.4) and (2.5) become:

$$(ineq.1) \qquad T_{surf}-T_{4,1} \geq 0$$

$$(ineq.2) \qquad T_{surf}-T_{4,M} \geq 0$$

$$(ineq.3) \qquad \Delta T_{surf}^2-(T_1(L)-T_M(L))^2 \geq 0$$

$$(ineq.4) \qquad B_{coke}-B_3^u-B_3^l-B_4^u-B_{EX} \geq 0$$

$$(ineq.5) \qquad B_2^u-\underline{B}_2^u \geq 0$$

$$\vdots$$

$$(ineq.14) \qquad \overline{B}_4^u-B_4^u \geq 0$$

Objective: (2.6)

$$(obj) \qquad a_2(B_2^u+B_2^l) + a_3(B_3^u+B_3^l) + a_4B_4^u = Min!$$

Figure 2.2 shows the unknowns and the equality constraints.

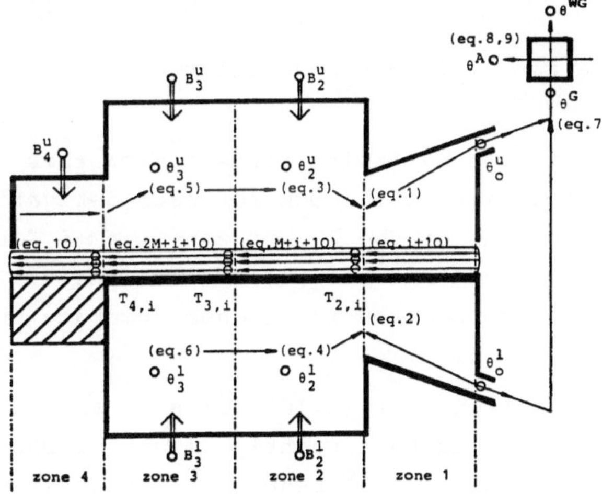

2.3   The Evaluation of the Restrictions

This chapter deals with the evaluation of the restrictions
and their Jacobian at a given approximation $x^{(k)}$ of the optimal
solution vector. The Hessian of the restrictions is not required
for the optimization algorithm of chapter 2.4. If the solutions
of the ODE's for the four zones and their derivatives are available, the restrictions and their Jacobian are easy to evaluate.
Thus, in the next sections we will discuss the solving of the
IVP's and the BVP.

2.3.1 Zone 1. Given the initial values $u(O)=u^{O}:=(\theta_{o}^{u},T_{A},..,T_{A},\theta_{o}^{l})$
we must compute the solution $u(L_1)$ of the ODE system (2.14),
(2.1o), (2.11), (2.12), (2.15), shortly

(2.16) $\quad \dfrac{\partial}{\partial x} u(x,p,u^{O}) = f(x,u,p)$

The notation $u(x,p,u^{O})$ points out that the solution depends on
both the initial value u^{O} and the parameters p of the right hand
side. At zone 1 p denotes the gas mass streams mg_2^{u} and mg_2^{l}, which
are functions of B. The derivatives with respect to B can be
calculated from those with respect to mg_2^{u} and mg_2^{l} using the chain
rule.

For the calculation of the derivatives $\dfrac{\partial u(L_1)}{\partial \theta_{o}^{u}}$, $\dfrac{\partial u(L_1)}{\partial \theta_{o}^{l}}$,
$\dfrac{\partial u(L_1)}{\partial mg_2^{u}}$ and $\dfrac{\partial u(L_1)}{\partial mg_2^{l}}$ we have two different possibilities:

- calculation of an approximation by finite differences: this
  requires one additional solution of the ODE for each derivative,
  and thus 4 additional solutions of the IVP of zone 1, M+4 for
  both the IVP of zone 2 and 3 and M+2 for the BVP of zone 4. If
  an order two approximation is required for higher accuracy the
  additional amount is twice as high.

- solving the sensitivity equations: If (2.16) holds in an open
  neighborhood of u^{O} and p and f is sufficiently smooth we can
  differentiate (2.16) with respect to u^{O} resp. p and change the

order of differentiation on the left hand side. The resulting
ODE's for the sensitivities are

$$(2.17) \quad \frac{\partial}{\partial x} \left(\frac{\partial}{\partial u^o} u(x,u^o,p) \right) = f'_u(x,u,p) \frac{\partial}{\partial u^o} u(x,u^o,p)$$

$$(2.18) \quad \frac{\partial}{\partial x} \left(\frac{\partial}{\partial p} u(x,u^o,p) \right) = f'_u(x,u,p) \frac{\partial}{\partial p} u(x,u^o,p) + f'_p(x,u,p).$$

The initial values for (2.17) are $\frac{\partial}{\partial u^o_j} u(o,u^o,p) = e_j$, $j \in \{1,M+2\}$,
where u^o_j is component j of the initial value u^o, those for
(2.18) are $\frac{\partial}{\partial p} u(o,u^o,p) = 0$ for $p = mg^u_2$ and $p = mg^1_2$.

Equations (2.16), (2.17), (2.18) together form a 5(M+2)-
dimensional IVP with a very special structure. The following
algorithm allows an efficient solving of this system by exploiting that structure: We integrate the heat equation with the
unconditionally stable Crank-Nicolson-method, which is the
trapezoidal rule for (2.16), and obtain the nonlinear system

$$(2.19) \quad u^+ - u = \frac{h}{2} [f(x,u,p) + f(x^+,u^+,p)]$$

for the approximation u^+ of $u(x^+) = u(x+h)$. u is an approximation
of u(x). Applying Newtons method to (2.19) we obtain the linear
equation

$$(2.2o) \quad [I - \frac{h}{2} f'_u(x^+,u^{+,k},p)] \Delta u^k = u - u^{+,k} + \frac{h}{2}[f(x,u,p) + f(x^+,u^{+,k},p)]$$

and we get the approximation $u^{+,k+1}$ of u^+ from $u^{+,k+1} = u^{+,k} + \Delta u^k$.
An initial approximation $u^{+,o}$ can be computed by an Euler predictor: $u^{+,o} = u + hf(x,u,p)$. The iteration continues until
$\|\Delta u^k\| < \varepsilon$ for some $\bar{k}$.

Before discussing the integration of (2.17) and (2.18) we
introduce the notation $\hat{u}(x) := \frac{\partial}{\partial u^o} u(x)$ and $\tilde{u}(x) := \frac{\partial}{\partial p} u(x)$,
again we write $\hat{u}$ and $\hat{u}^+$ for the approximations of $\hat{u}(x)$ and
$\hat{u}(x^+)$. The trapezoidal rule for (2.17) gives

$$\hat{u}^{+} - \hat{u} = \frac{h}{2}[f_u'(x,u,p)\hat{u} + f_u'(x^{+},u^{+},p)\hat{u}^{+}],$$

which is equivalent to

$$(2.21) \quad [I - \frac{h}{2}\,f_u'(x^{+},u^{+},p)]\hat{u}^{+} = \hat{u} + \frac{h}{2}\,f_u'(x,u,p)\hat{u}.$$

For (2.18) we get the linear system

$$(2.22) \quad [I - \frac{h}{2}\,f_u'(x^{+},u^{+},p)]\tilde{u}^{+} = \tilde{u} + \frac{h}{2}\,[f_u'(x,u,p)\tilde{u} + f_p'(x,u,p) +$$
$$+ f_p'(x^{+},u^{+},p)]$$

The system matrices of (2.21) and (2.22) are identical and have
the same form as that of (2.2o), but they are evaluated at
$u = u^{+}$ instead of $u = u^{+,k}$. As $u^{+} = u^{+,\bar{k}+1}$ and $\|u^{+,\bar{k}+1} - u^{+,\bar{k}}\| =$
$= \|\Delta u^{\bar{k}}\| < \varepsilon$ and f_u' is continuous, the difference of the two
matrices is small and we can use the Jacobian of (2.2o) for
$k = \bar{k}$ (last Newton step) to solve (2.21) and (2.22). In addition,
we even have already a factorization of the matrix after solving
(2.2o). Thus the integration of the sensitivity equations only
demands the computation of the right hand sides of systems (2.21)
and (2.22) ($f_u'(x,u,p)$ is known from the former integration step,
also $f_p'(x,u,p)$, $f_p'(x^{+},u^{+},p)$ is to evaluate) and the solving of
two triangular linear systems for each sensitivity. The
additional amount for the sensitivity calculation is thus very
small.

If a very high accuracy of the sensitivities is necessary,
an additional evaluation of the Jacobian at u^{+} is required. The
sensitivities calculated by this method are not only approximations of the solutions of the sensitivity equations, but they
are also the sensitivities of the solution of the discretization
(2.19) of the ODE, and thus exact (if (2.19) is solved exactly).

The method is completed by a stepsize control and local
linear extrapolation to increase the order of convergence from
two to four. As the extrapolation is linear, it does not affect

the exactness of the sensitivities, but it destroys the absolute
stability of the trapezoidal rule. A numerical eigenvalue
analysis of f_u' showed that for $M \leq 11$ the stability condition
$hq \geq -25,8$ ($q = \min\{q_i | q_i$ eigenvalue of $f_u'\}$, h stepsize) was
never hurt, while for $M = 13$ some of the integration steps at
the end of zone 1 hurt this condition (note: $|q| = O(M^2)$). During
numerical tests no loss of stability could be noticed. Therefore, a stepsize restriction, which could be realized easily
because of good eigenvalue estimates, did not seem necessary.

2.3.2 <u>Zones 2 and 3.</u> The solving of the IVP's for zones 2 and 3
and the sensitivity calculations with respect to the initial
values $T_{2,i}$ resp. $T_{3,i}$ and the parameters θ_2^u, θ_2^l, θ_3^u and θ_3^l are
performed with the method described in section 2.3.1. The
integral terms of (eg.3), (eg.4), (eg.5) and (eg.6) are transformed into ODE's, f.e.

$$\frac{d}{dx} h(x) = 1c_2^u(x)(\theta_2^{u^4} - T_1^4)$$

with $h(L_2) = O$. Then we have $h(L_3) = 1\int_{L_2}^{L_3} c_2^u(x)(\theta_2^{u^4} - T_1^4)dx$, which

is the second integral term of (eg.3). The additional ODE's,
four for both zones 2 and 3, are solved simultanously with the
heat equation.

2.3.3 <u>Zone 4.</u> The BVP describing zone 4 consists of the ODE's
(2.13), (2.1o), (2.11), (2.12) and the separated boundary
conditions $\theta_4^u(L) = \bar{\theta}_4^u$ and $T_i(L_3) = T_{4,i}$ - shortly $u' = f(x,u)$,
$u_1(L) = \bar{u}_1$, $u_i(L_3) = \bar{u}_i$ for $i = 2,\ldots,M+1$. Because of its
physical properties (heat equation, countercurrent flue gas
with low mass stream near $x = L$) the system is numerically very
unstable.

Efficient methods for solving BVP's are multiple
shooting methods MSM [7]. By choosing a sufficiently fine subdivision $\{x_j\}_{j=0,\ldots N_j}$ of the intervall $[L_3,L]$, the method gets

stabilized. As usual we introduce unknowns $S_j \sim u(x_j)$ for $j = 0,\ldots,N_j$ and define the initial value problems

$$(\text{IVP}_j) \quad u_j' = f(x,u_j) \qquad x \in [x_j,x_{j+1}]$$

$$u_j(x_j) = S_j \qquad\qquad \text{for } j = 0,\ldots N_j-1$$

As our subdivision is very fine ($h = x_{j+1}-x_j = \dfrac{L-L_3}{N_j}$), (IVP_j) can be solved approximately by one step of the trapezoidal rule:

$$(2.23) \quad u_j(x_{j+1}) \sim t_j = S_j + \frac{h}{2}\left[f(x_j,S_j) + f(x_{j+1},t_j)\right].$$

The discretization error is further reduced by global extrapolation, which does not destroy the stability properties like local extrapolation does.

If the solutions u_j of (IVP_j) are continuous at x_j and if $u_0(L_3) = S_0$ and $u_{N_j-1}(L)$ satisfy the boundary conditions, they also solve the BVP. The continuity condition for the approximate solution is

$$(2.24) \quad t_j = S_{j+1} \quad \text{for } j = 0,\ldots,N_j-1$$

and the boundary conditions are

$$(2.25) \quad (S_0)_i = \bar{u}_i \quad \text{for } i = 2,\ldots,M+1 \text{ and } (t_{N_j-1})_1 = \bar{u}_1$$

System (2.23), (2.24), (2.25) is a set of $(M+1)(2N_j+1)$ equations for the unknowns S_j and t_j. The simple form of (2.24) allows the elimination of the t_j's, the resulting system

$$(2.26) \quad S_{j+1} = S_j + \frac{h}{2}\left[f(x_j,S_j)+f(x_{j+1},S_{j+1})\right] \quad \text{for } j = 0,\ldots N_j-1$$

$$(2.27) \quad (S_0)_i = \bar{u}_i \text{ for } i = 2,\ldots,M+1 \text{ and } (S_{N_j})_1 = \bar{u}_1$$

is highly sparse. The Jacobian is a band matrix with one sub- and M+2 superdiagonals and one additional nonzero in the left lower edge. The system, shortly $F(S,p) = 0$, is solved by Newtons method:

$$(2.28) \quad F_S'(S^k,p)\Delta S^k + F(S^k) = 0$$

$$S^{k+1} = S^k + \Delta S^k$$

The sensitivities $\frac{dS}{dp}$ of the numerical solution of the BVP with respect to $T_{4,i}$, B_4^u and θ^A, which are the parameters p of the system F, are computed from the following system:

$$(2.29) \quad 0 = \frac{d}{dp} F(S,p) = F_S'(S,p) \frac{dS}{dp} + F_p'(S,p)$$

The Jacobian F_S' is, at least approximately, known from Newtons method (2.28), F_p' is easy to compute. We can even use the decomposition of F_S' from the solving of (2.28), and so the sensitivity calculation again is very cheap.

2.4   Simulation and Optimization, Examples

2.4.1 Simulation. As a first step we implemented a program for the simulation of the reheating process. It allowed us a first check of the model and the numerical methods for the function evaluation. In addition of equation (eg.1),...,(eg.3M+1o) we took the four conditions

$$(2.3o) \quad w_3^u B_2^u - w_2^u B_3^u = 0, \qquad w_3^u B_2^l - w_2^l B_3^u = 0,$$

$$w_3^u B_3^l - w_3^l B_3^u = 0, \qquad w_3^u B_4^u - w_4^u B_3^u = 0$$

to obtain a system of (3M+14) equations for the (3M+14) un- knowns. Conditions (2.3o) say that the total full demand is distributed to the zones according to the weights w, which must be fixed by the user.

The system is solved by Newton's method, damping is not
necessary. The resulting linear systems are solved by the
Gaussian algorithm.

2.4.2 <u>Optimization.</u> The optimization problem of chapter 2.2.4
has the standard form

$$f(x) = \text{Min!}$$
$$(2.31) \quad g_j(x) = 0 \quad \text{for } j = 1,\ldots,m1$$
$$g_j(x) \geq 0 \quad \text{for } j = m1+1,\ldots,m$$

We have 3M+14 unknowns, 3M+1o equality and 14 inequality
constraints. Although the methods of chapter 2.3 allow an efficient evaluation of the restrictions and their Jacobian, the
function evaluation is still the main computational effort for
the optimization. Thus we decided on a sequential quadratic
programming method for the optimization as these methods require
relativly few function evaluations. Newton's method is only
possible in its discretized form, for which we would need a lot
of additional function evaluations. Therefore, we use a Quasi
Newton Method, which has good convergence properties - for
instance because of the low dimension (between zero and four) of
the tangent cone of the problem. In particular, we took the algorithm of K.Schittkowski [4] and adapted it to our special problem.

Starting at some approximation $x^{(k)}$, the method defines
the quadratic subproblem

$$d^T Bd + c^T d = \text{Min!}$$
$$(2.32) \quad a_j^T d + b_j = 0 \quad j = 1,\ldots,m1$$
$$a_j^T d + b_j \geq 0 \quad j = m1+1,\ldots,m$$

with $a_j = \nabla_x g_j(x^{(k)})$, $b_j = g_j(x^{(k)})$ and $c = \nabla_x f(x^{(k)})$. B is a
Quasi Newton approximation of the Hessian of the Lagrangian. The
solution d gives a search direction for x.

But as (2.32) may not have a feasible point, an additional parameter δ ($0 \leq \delta \leq 1$) is introduced:

$$
\begin{aligned}
d^T B d + c^T d + \rho \delta^2 &= \text{Min!} \\
a_j^T d + (1-\delta) b_j &= 0 && j = 1,\ldots,m1 \\
a_j^T d + (1-\delta) b_j &\geq 0 && j \in \{m1+1,\ldots,m \mid b_j \leq 0\} \\
a_j^T d + b_j &\geq 0 && j \in \{m1+1,\ldots,m \mid b_j > 0\}
\end{aligned}
$$

(2.33)

For $\delta = 1$ $d = 0$ is feasible, for $\delta = 0$ (2.33) is equivalent to (2.32). The quadratic problem is solved by an algorithm of Gill and Murray [1] which uses a Q-R-factorization of the Jacobian. If no feasible point for (2.32) exists, $\delta = 0$ is not possible, but for $\delta \leq \bar{\delta}$ the solution of (2.33) gives a search direction for (2.31).

The line search is performed along $x^{(k)} + \alpha d$ for $\alpha > 0$. The algorithm of Schittkowski uses an augmented Lagrangian line search function (Lagrangian plus penalty terms $r_j g_j^2(x)$).

The algorithm converges globally and has a local super-linear order of convergence. Our tests were satisfying: In most cases they required less than 1o function evaluations. For some test problems (e.g. high furnace capacity: v high) the algorithm had troubles with finding a feasible point. To cope with this problem we dropped some of the inequality restrictions (e.g. (ineq.1),(ineq.2), (ineq.3)) until a feasible point of this easier problem was found. Starting from this point, we tried to find a solution of the original problem.

2.4.3 <u>Examples.</u> Fig. 2.3 shows the results of a cold charging simulation. The heating time is 2 hours, and thus 313 686 kg/h steel are heated from $3o^{\circ}C$ to $125o^{\circ}C$. T_1 is the upper surface temperature, T_b the bulk temperature and T_M the lower surface temperature. θ^u and θ^l are the flue gas temperatures. The given fuel stream distribution was 19 % resp. 29 % for zone 2 (upper resp. lower furnace part), 21 % resp. 28 % for zone 3 and 3 %

for zone 4. The total energy demand is $132,58.1o^6$ W, and the solution violates both the restriction for the upper and the lower surface temperature at L_3 and the restriction for the total coke gas demand. The optimal solution for the same example, which violates no restriction, is shown in Fig. 2.4. The energy demand reduces to $13o,86.1o^6$ W, the difference is about 1 %. For the same production rate, but hot charging ($5oo^o$ C $\rightarrow$ $125o^o$ C), the energy demand is $82,61.1o^6$ W, the optimal solution is shown in Fig. 2.5.

The algorithm required 9 function evaluations for both of the optimization problems and 6 function evaluations for the simulation.

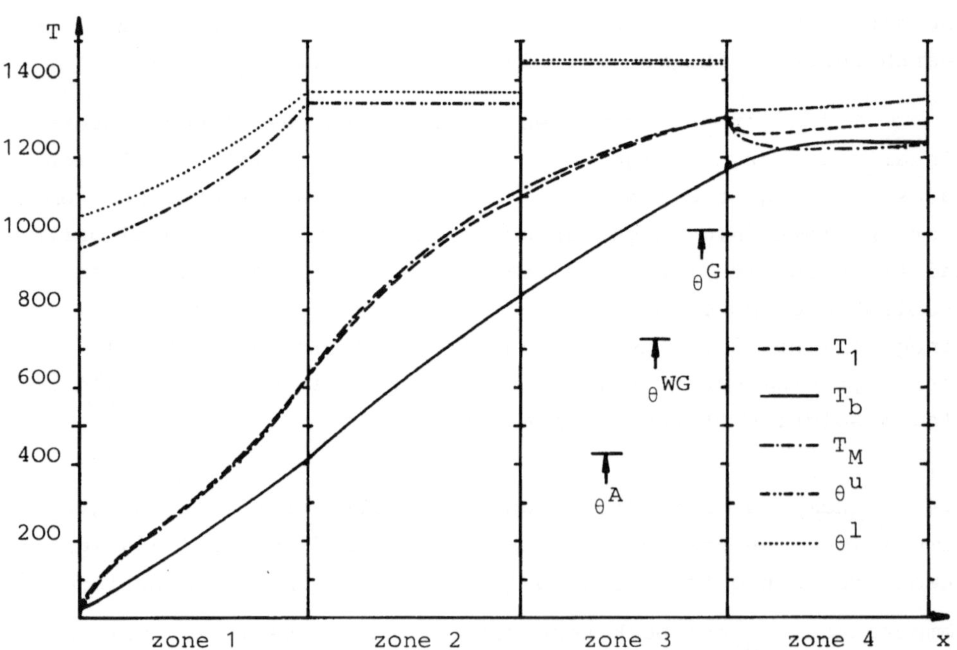

Fig. 2.3 cold charging simulation

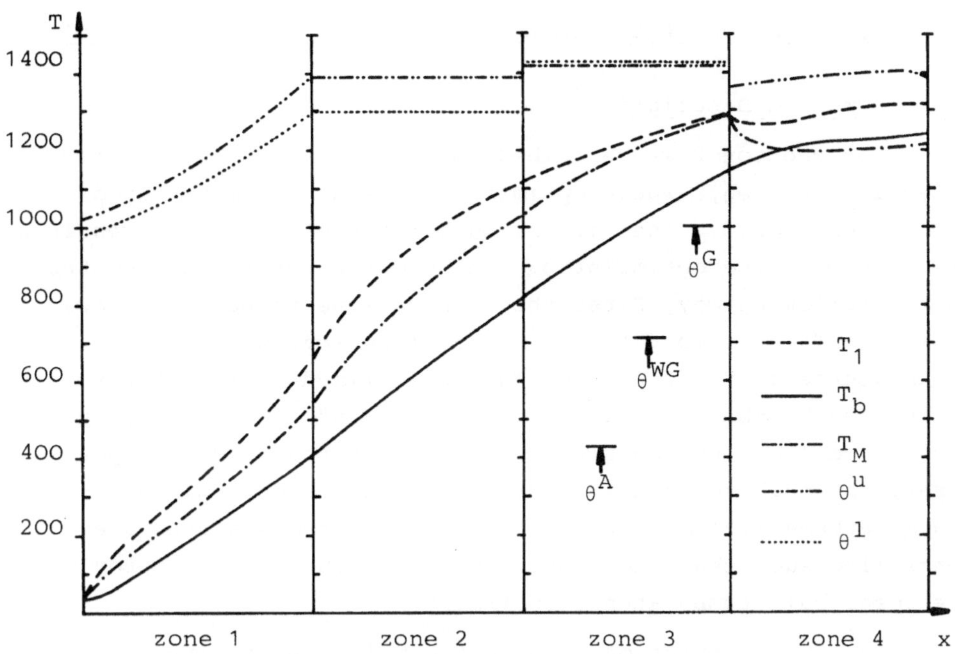

Fig. 2.4 cold charging optimization

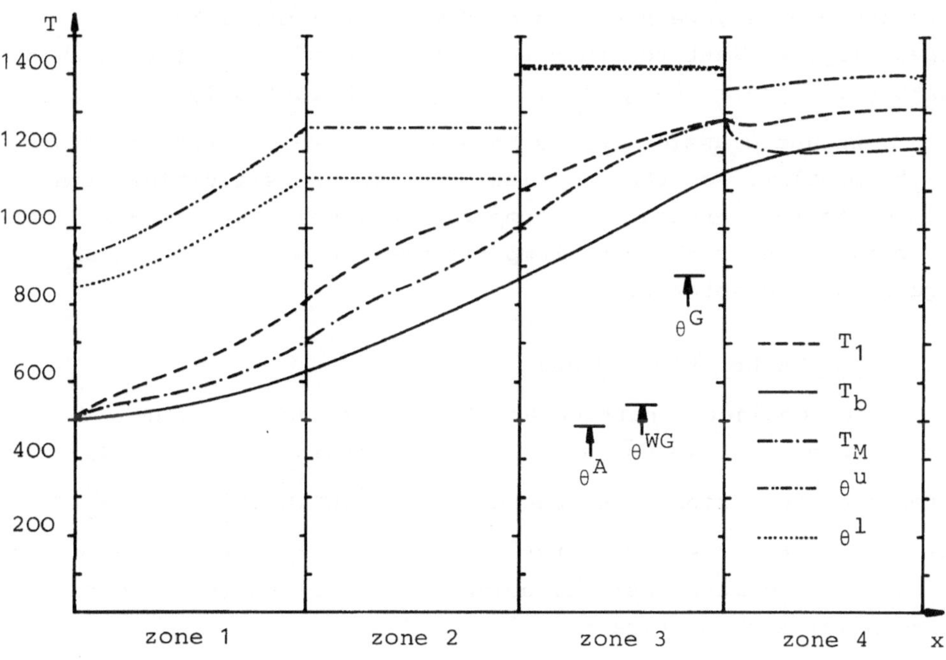

Fig. 2.5 hot charging optimization

3   Calculation of the Heat Distribution - Furnace Control

3.1   Problem Description

In section 2 we described how to determine the optimal
reheating strategy, where optimization means minimization of
energy resp. costs. The "ideal" bulk temperature curves resulting
from the off-line optimization are used for the furnace control
in the following way. First the furnace zone temperatures are
measured. These temperatures are not identical with the
temperatures of the flue gas. They are "mixed" temperatures
influenced by the flue gas, by the flames of the burners, by the
furnace wall and by the slabs. Given the furnace zone temperatures, the bulk temperatures of the slabs are determined on-line
via a mathematical model. The furnace zone temperatures are
controlled such that the resulting bulk temperatures approximate
the ideal bulk temperatures optimally.

As the computation of both the bulk temperatures and the
zone temperatures must be performed on-line three dimensional
models do not seem appropriate. But in order to get more insight
into the process, we use a three dimensional model here.
Especially, we want to study the influence of the skids as skid
marks deteriorate the quality of steel substantially.

We use classical FE-techniques and exploit the structure
of the problem. For the calculation of the sensitivities with
respect to the furnace zone temperatures, which are required for
the correction of the zone temperatures, we use similar techniques as in chapter 2.3.

3.2   The Mathematical Model

We consider a particular slab being pushed through
the furnace, i.e. we use a slab fixed coordinate system. We
determine those zone temperatures $W = (W_1^u, W_1^l, W_2^u, W_2^l, W_3^u, W_3^l, W_4^u)$,
which give the bulk temperature curve closest to the ideal curve.
We work in 3 dimensions: the height of the slabs (y), their
breadth (z) and the time t.

At first we reduce the size of the problem by symmetry considerations.

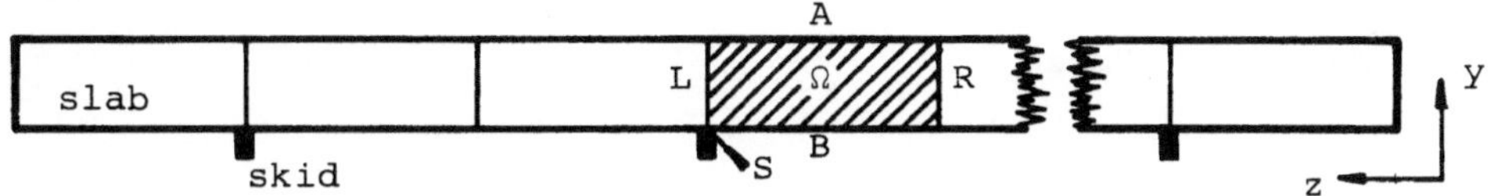

Fig. 3.1   slab

Under the assumption $\left.\dfrac{\partial T}{\partial z}\right|_{L\,U\,R} = 0$ it suffices to consider Ω. We get the following model for the heat distribution

$$(3.1) \quad \underbrace{c_p(T)\rho}_{=:\ r(T)}\ \frac{\partial T}{\partial t} = \operatorname{div}[\lambda(T)\ \operatorname{grad} T] \qquad (y,z) \in \Omega,\ t \in (0,t_E)$$

$$(3.2) \quad \left. T(y,z,t)\right|_{t=0} = T_A$$

$$(3.3) \quad \lambda(T)\frac{\partial T}{\partial n}(y,z,t)=R(y,z,t,T)=\begin{cases} 0 & L\,U\,R,\text{zones } 1\text{-}4 \\[2mm] h_i^u(W_i^u-T)+r_i^u(W_i^{u\,4}-T^4) & A,\text{zones } 1\text{-}4 \\[2mm] h_i^l(W_i^l-T)+r_i^l(W_i^{l\,4}-T^4) & B,\text{zones } 1\text{-}3 \\[2mm] k^S(T^C-T) & S,\text{zones } 1\text{-}3 \\[2mm] k^H(T^H-T) & B\,U\,S,\text{zone } 4 \end{cases}$$

h_i^u and h_i^l are the convectivity constants at zone i, r_i^u and r_i^l the radiation constants, k^S resp. k^H the heat exchange coefficients slab-skids-cooling water resp. slab-soaking hearth. For the approximation of the ideal bulk temperature $T_{opt}(t)$ by the bulk temperature $T(\bar{y},\bar{z},t)$ in $L_2[0,t_E]$ we obtain the objective

$$(3.4) \quad f(W) = \int_o^{t_E} [T(\bar{y},\bar{z},t,W)-T_{opt}(t)]^2 dt = \text{Min!}$$

The control W must satisfy the (technical) box constraints

$$(3.5) \quad W_{min} \leq W \leq W_{max}.$$

3.3   The Determination of the Heat Distribution

We assume W to be given.

3.3.1 Time Discretization: Trapezoidal Rule.

We introduce the time discretization $t_o := 0$, $t_{n+1} := t_n + \Delta t_n$ with Δt_n variable and write $T_n(y,z)$ for the numerical approximation of $T(y,z,t_n)$. To preserve stability we use the trapezoidal rule:

$$(3.6) \quad \frac{(T_{n+1} - T_n)(y,z)}{\Delta t_n} = \frac{1}{2} \left[\frac{D(T_n)}{r(T_n)} + \frac{D(T_{n+1})}{r(T_{n+1})} \right]$$

with $D(T) := \mathrm{div}(\lambda(T)\mathrm{grad}\, T) = \lambda'(T)\nabla T^2 + \lambda(T)\Delta T$. (3.6) is a non-linear elliptic differential equation for T_{n+1}.

3.3.2 Linearization.

We solve (3.6) by a predictor-corrector technique, using a trivial predictor and a Newton corrector (Fig. 3.4):

predictor P: $\tilde{T}_{n+1} := T_n$

corrector C: $T_{n+1} := \tilde{T}_{n+1} + u$, where $u = u(y,z)$ satisfies

$$(3.7) \quad \frac{u}{\Delta t_n} = \frac{1}{2} \left[\frac{D(T_n)}{r(T_n)} + \frac{D(T_n)}{r(T_n)} + \frac{D'(T_n)u}{r(T_n)} - \frac{D(T_n)r'(T_n)u}{r^2(T_n)} \right].$$

(3.7) is the linearization of (3.6) at $\tilde{T}_{n+1} = T_n$. We have

$$D'(T)u = \lambda''(T)\nabla T^2 u + 2\lambda'(T)\nabla T\nabla u + \lambda'(T)\Delta Tu + \lambda(T)\Delta u.$$

For the linearized boundary condition at boundary A, for instance, we get:

$$(3.8) \quad \lambda(T)\frac{\partial u}{\partial n} = -[h_i^u + 4r_i^u T_n^3 + \lambda'(T_n)\frac{\partial T_n}{\partial n}]u,$$

those for B,S,L and R are analogous. This model holds inside each zone. Some care is necessary when passing from one zone to the next one as we have jumps in the right hand side of (3.3) at these points (see [2], [3]).

3.3.3 <u>Transformation to Standard Form.</u> Multiplying (3.7) with $2r(T_n)$ and applying the transformation

$$v = v(y,z) := \lambda(T_n(y,z)).u(y,z)$$

to the system (3.7), (3.8), we get the standard form

$$(3.9) \quad -\Delta v + a(y,z)v = b(y,z) \qquad (y,z) \in \Omega$$

$$(3.10) \quad \frac{\partial v}{\partial n} = \alpha(y,z).v \qquad (y,z) \in \partial\Omega$$

with $a := \dfrac{2}{\Delta t_n}\dfrac{r(T_n)}{\lambda(T_n)} + \dfrac{r'(T_n)\lambda'(T_n)}{r(T_n)\lambda(T_n)} \nabla T_n^2 + \dfrac{r'(T_n)}{r(T_n)} \Delta T_n,$

$b := 2[\lambda'(T_n) \nabla T_n^2 + \lambda(T_n)\Delta T_n]$ and

$\alpha := \dfrac{-1}{\lambda(T_n)} [h_i^u + 4r_i^u T_n^3 + \lambda'(T_n)\dfrac{\partial T_n}{\partial n}]$ for boundary A, e.g.

3.3.4 <u>Galerkin.</u> We use the following ansatz for $v(y,z)$:

$$v(y,z) = \sum_{k=1}^{n} c_k N_k(y,z)$$

with finitly many elements N_k. Then the method of Galerkin (checking equation (3.9), (3.10) in the weak sense with respect to each ansatz function N_j) gives the linear system

$$\iint_\Omega -\Delta\left(\sum_{k=1}^n c_k N_k\right)N_j \ dydz + \iint_\Omega a(y,z)\left(\sum_{k=1}^n c_k N_k\right)N_j \ dydz =$$

$$= \iint_\Omega b(y,z)N_j \ dydz$$

For eliminating the second derivatives, we apply Greens theorem

$$\iint_\Omega \Delta v_1 v_2 \ dydz = -\iint_\Omega \nabla v_1 \nabla v_2 \ dydz + \int_{\partial\Omega} \frac{\partial v_1}{\partial n} v_2 \ ds$$

to all the terms containing the Laplace operator and get:

$$(3.11) \quad \sum_{k=1}^n c_k \Big\{ \iint_\Omega \nabla N_k \nabla N_j - \frac{r'}{r} \nabla T (\nabla N_k N_j + N_k \nabla N_j) +$$

$$+\left[\frac{2}{\Delta t_n} \frac{r}{\lambda} + \left(\frac{r'\lambda'}{r\lambda} - \left(\frac{r'}{r}\right)'\right)\nabla T\nabla T\right] N_k N_j \ dydz +$$

$$+ \int_{\partial\Omega} \left[\frac{r'}{r} \frac{R}{\lambda} -\alpha\right] N_k N_j \ ds \Big\} = -2\iint_\Omega \lambda \nabla T \nabla N_j \ dydz + 2 \int_{\partial\Omega} R \ N_j \ ds$$

$$\text{for } j = 1,\ldots,n.$$

(3.11) is a positive definit linear equation for the n unknowns c_k. For a proper choice of the ansatz functions N_k and for sufficiently large n $\sum_{k=1}^n c_k N_k$ is a good approximation of v and $T_n + v$ approximates T_{n+1}. Integration in (3.11) is performed by Gaussian quadrature formula to gain accuracy.

3.1.5 <u>Finite Element Techniques.</u> We chose the finite element technique for generating the ansatz functions. Ω is divided into 276 triangles and each ansatz function N_k is a piecewise linear function which is zero at all knots but one, where it has the value 1. A knot is a vertex of a triangle; there are 168 knots. This particular form of N_k makes the evaluation of the integrals of (3.11) very easy. The system matrix of (3.11) is very sparse, because each integral contains a product of N_k resp. ∇N_k and N_j

resp. ∇N_j, which is zero for most of the pairs

$(k,j) \in \{1,\ldots,n\} \times \{1,\ldots,n\}$. As the structure of the matrix depends on the numbering of the knots, we used the algorithm of Cuthill-McKee (see [5]) to reduce the band-width. Fig. 3.2 shows the triangulation - near the skid it is finer to get a good picture of the skid marks - and the numbering of the knots resulting from the Cuthill-McKee algorithm. The band-width of the matrix is m = 12, and thus we need an array of n(m+1) = 168 x 13 = 3588

elements instead of 168^2 = 28224 elements for its storage. The amount of work for solving the linear equation is about $n(m^2+3m)/2$ operations.

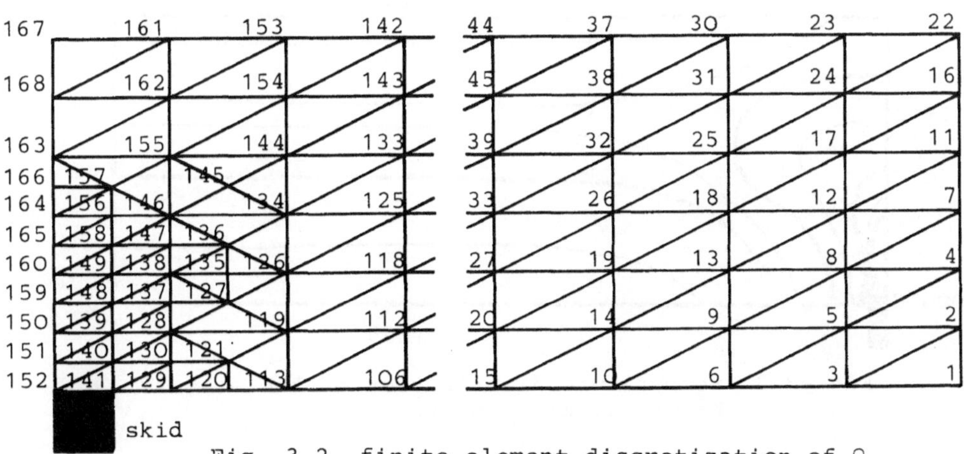

Fig. 3.2 finite element discretization of Ω

3.3.6 Stepsize Control. The accuracy of the results strongly depends on the time stepsize Δt_n. Thus we used the stepsize control proposed in [6], which gave good results. But as the integration has to be performed with both stepsizes Δt_n and $\frac{1}{2} \Delta t_n$, the numerical amount is high. So when we found out that the stepsize sequences are almost the same for all our test examples, we took this typical stepsize sequence and could so reduce the numerical amount drastically with only a small loss of accuracy. On an IBM 3o31 a stepsize controlled solution of our problem needs 17.64 min whereas for a mostly constant stepsize of 2oo [sec] the computation was finished after 4.o5 min.

The maximal difference of the solutions is less than 1 %.

In addition, for the optimization the fixed step size
sequence with partially constant stepsizes gave another substantial reduction of the numerical amount, see chapter 3.4.

3.3.7 <u>Numerical Results: Temperature Distribution</u>. The following
pictures give the temperature distribution of Ω at the end of
zones 3 and 4 (before and after the soaking hearth)

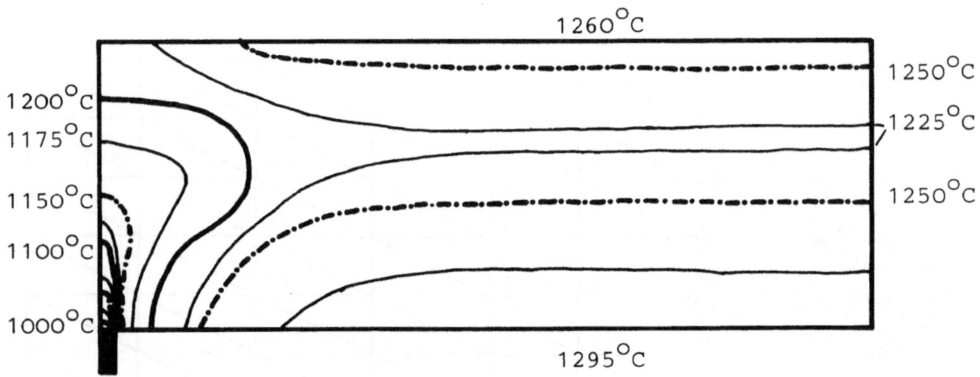

Fig. 3.3   temperature distribution of the slab (Ω)
          before soaking hearth

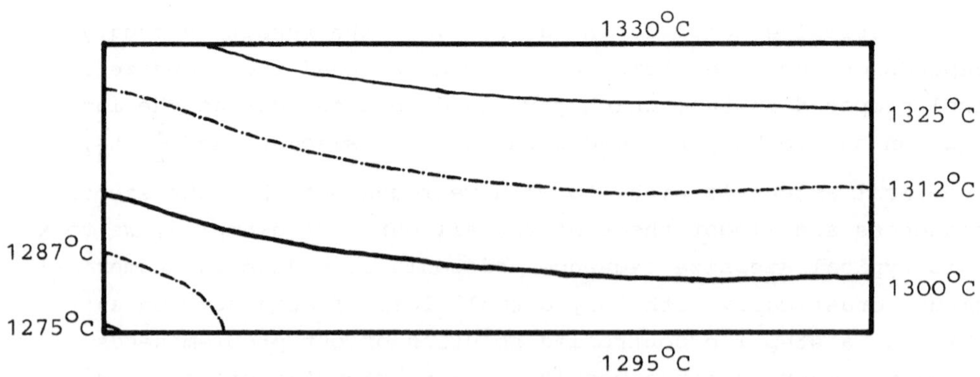

Fig. 3.4   temperature distribution of the slab (Ω)
          after soaking hearth

3.4   Optimization

In section 3.3 we described how to determine T in dependence of W. Now we want to solve

$$f(T(W)) = \text{Min!} \qquad W_{min} \leq W \leq W_{max}, \ W \in \mathbb{R}^7$$

As each function evaluation is very costly because of the computation of T, we decided on a Quasi Newton procedure using the BFGS-formula for the approximation of the Hessian. For the determination of the gradients we use a technique similar to that of section 2.3.

3.4.1   Determination of the Gradients. From (3.4) we get

$$(3.12) \quad \frac{\partial f}{\partial W}(T(W)) = \frac{\partial f}{\partial T}\frac{\partial T}{\partial W} = \underbrace{\int_{o}^{t_E} 2[T(x,y,T,W)-T_{opt}(t)]}_{=\,:\,K} \underbrace{\frac{\partial T}{\partial W}}_{=:U} dt$$

To determine U we differentiate the PDE (3.1), (3.2), (3.3) and change the order of differentiation:

$$(3.13) \quad \frac{\partial U}{\partial t} = - \frac{r'(T)}{r^2(T)} D(T).U + \frac{D'(T)}{r(T)} U$$

$$(3.14) \quad U(y,z,t)\Big|_{t=0} = O$$

$$(3.15) \quad \lambda'(T)\frac{\partial T}{\partial n} U + \lambda(T)\frac{\partial U}{\partial n} = R'_T(T,W)U + R'_W(T,W)$$

The seven parabolic problems (3.13), (3.14), (3.15) are solved by exactly the same techniques - time discretization, transformation $V_n(y,z) = \lambda(T_n)U_n(y,z)$ - as problem (3.1), (3.2), (3.3). There is no linearization as the problems are already linear. At each time step we have to solve

$$(3.16) \quad -\Delta V_n + a(y,z)V_n = n(y,z)$$

$$(3.17) \quad \frac{\partial V_n}{\partial n} = \alpha(y,z)V_n + \beta(y,z)$$

114

Those parts of the system containing the unknown V_n are exactly
the same as those of system (3.9), (3.1o). The only difference
is that in (3.9), (3.1o) the functions $a(y,z)$ and $\alpha(y,z)$, which
depend on T, are to evaluate at $\tilde{T}_{n+1} = T_n$, while for system
(3.16), (3.17) we require the functions at T_{n+1}. But as
$T_{n+1} = \tilde{T}_{n+2}$, the system matrix of (3.16), (3.17) - after
application of Galerkins method and finite element discretization
- is exactly the same as that of (3.11) at the next time step,
provided that $\Delta t_n = \Delta t_{n+1}$. Even the amount of work for the evaluation of the right hand sides can be substantially reduced by a
similar consideration. Assume, we know (T_n, U_n) and want to
determine (T_{n+1}, U_{n+1}) inside a zone (see also Fig. 3.4)

Step 1: Determine T_{n+1} by solving (3.11). The system matrix at
$\tilde{T}_{n+1} = T_n$ and its Cholesky factorization are already
known

Step 2: Determine the Jacobian at T_{n+1}, which is the system
matrix for both the systems for V_{n+1} and the system for
T_{n+2}.

Step 3: Determine the right hand sides for the systems for
V_{n+1} and solve the equations. Back transformation gives
U_{n+1}.

Some care is necessary when passing from one zone to the next
one, for details see [2]. For changing stepsizes $(\Delta t_{n+1} \neq \Delta t_n)$
only little additional amount of work is necessary.

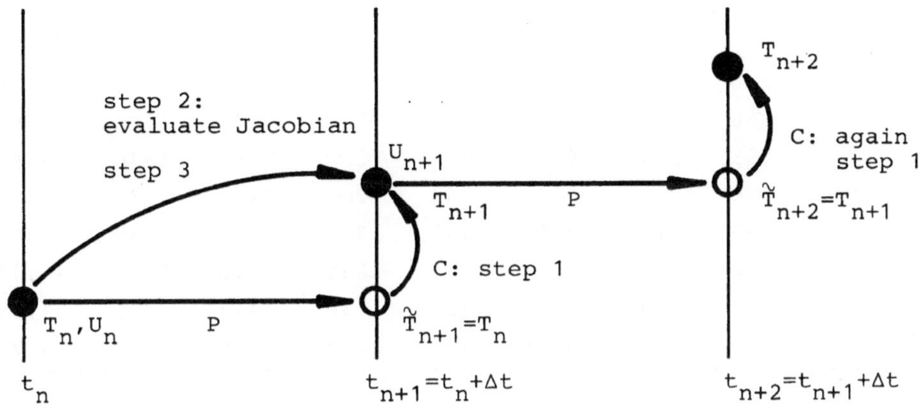

Fig. 3.5   calculation of T_{n+1} and U_{n+1}

3.4.2 <u>Minimization of f(W).</u> We determine $f(W)$ and $\frac{\partial f}{\partial W}$ (W) by the trapezoidal rule using just those points where T and U were determined. Then we use the following well known algorithm

Start: $f(W_0)$, $\nabla f(W_0) = : g_0$, $B_0 = \beta I$, k: = 0

S1: $B_k S_k = -g_k \implies S_k$

S2: $f(W_k + \alpha S_k) = $ Min, $W_{min} \leq W \leq W_{max} \implies \alpha_k$

$W_{k+1} = W_k + \alpha_k S_k$ (Line search is performed by quadratic interpolation)

S3: W_{k+1} optimal $\implies$ stop

S4: B_{k+1} by BFGS-Formula, k = k+1, GOTO S1

3.4.3 <u>Numerical Results.</u> Depending on the charging temperature, the quality of steel, the size of the slabs and the pushing velocity v, there are different optimal curves $T_{opt}(t) = T_{opt}(\frac{x}{v})$ for the bulk temperature, see chapter 2.4.3. We will present the approximation of one typical cold charging optimal curve. In the following table the results are rounded. $\Delta T = \sqrt{\frac{f(W)}{t_E}}$ gives the mean difference of the calculated and the optimal bulk temperature.

Iteration	W_1^u	W_1^l	W_2^u	W_2^l	W_3^u	W_3^l	W_4^u	ΔT
1	6oo	55o	1ooo	1o5o	13oo	135o	135o	129,96
2	566	521	922	961	1196	1242	133o	58,93
3	563	519	924	965	12o9	126o	1336	58,97
line search	656	52o	923	963	12o2	1251	1335	58,7o
4	555	512	915	957	121o	1263	1339	54,81
5	52o	484	886	933	1228	1296	1351	41,13
6	487	459	855	9o8	1236	1318	136o	31,69
7	45o	429	816	877	1236	1336	1368	28,97

Acknowledgements:

The authors are indebted to Dipl.-Ing.B.Lindorfer (VOEST-ALPINE AG)
and Dr.W.Zulehner for their contributions to modelling and to
Dr.H.Gfrerer for his advice concerning optimization.
Financial support was given by the National Banking Foundations
and the VOEST-ALPINE AG.

References:

[1] P.E.Gill, W.Murray, "Numerically stable methods for quadra-
     tic programming", Math.Prog. 14, 1978, pp.349-372

[2] R.Hubmer, "Optimierung eines Stoßofens: Sollkurvenapproxi-
     mation unter Anwendung der Methode der Finiten Elemente",
     diploma thesis 1986, Johannes-Kepler-University, Linz

[3] M.Lindner, "Hierarchische Formfunktionen angewandt auf ein
     parabolisches Anfangsrandwertproblem", diploma thesis 1987,
     Johannes-Kepler-University Linz

[4] K.Schittkowski, "On the convergence of a sequential qua-
     dratic programming method with an augmented Lagrangian
     line search function", Optimization 14, 1983, pp.197-216.

[5] H.R.Schwarz, "Methode der finiten Elemente", Teubner,
     Stuttgart 1984

[6] K.Solchenbach, K.Stüben, U.Trottenberg, K.Witsch, "Efficient
     solution of a nonlinear heat conduction problem by multigrid
     methods", Sonderforschungsbereich 72, Universität Bonn

[7] J.Stoer, R.Bulirsch, "Introduction to Numerical Analysis",
     Springer 198o

ON THE DESIGN OF THE VOLUTE OF A CENTRIFUGAL PUMP

W.Zulehner

1   Introduction

This paper contains the results of a joint research project by M.Kowalik (Firma Ochsner & Sohn, a company in Linz
manufacturing pumps), E.Zarzer and W.Zulehner (University of
Linz), see [1].

Figure 1 is a sketch of a centrifugal pump, for details
see e.g. [4], [5], [7]. Arrows mark the direction of flow.

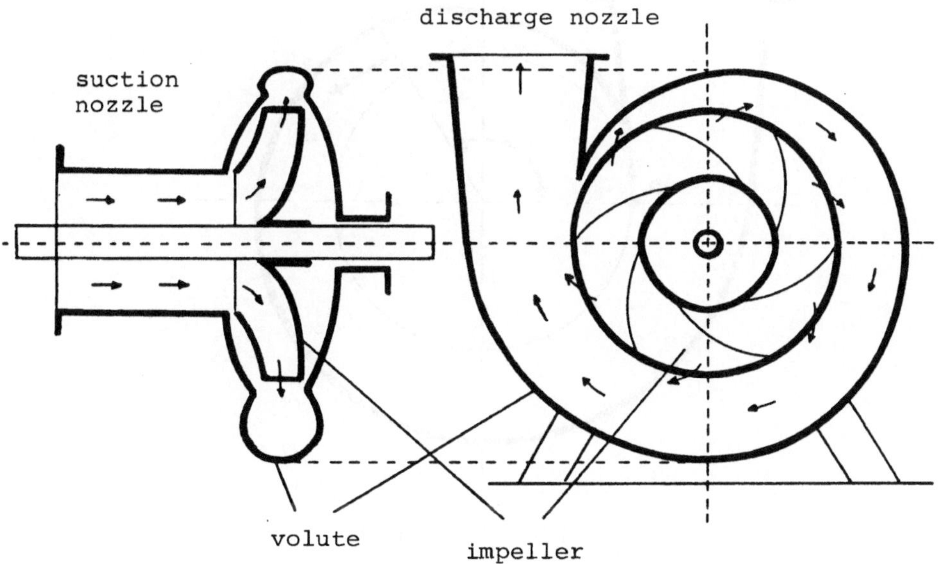

Fig. 1   Centrifugal pump

Liquid enters the pump through a suction nozzle and is brought into circular motion by an impeller. Due to centrifugal forces, it leaves the impeller at a higher pressure and higher velocity and is led into a casing channel called the volute. Finally, liquid leaves the pump through a discharge nozzle. The problem is how to design the volute in order to obtain a pump of high efficiency. (Efficiency is defined as ratio of pump energy output to energy input.)

2   <u>The engineering model</u>

First of all, several geometric assumptions are made, for notation see Fig. 2 and Fig. 3.

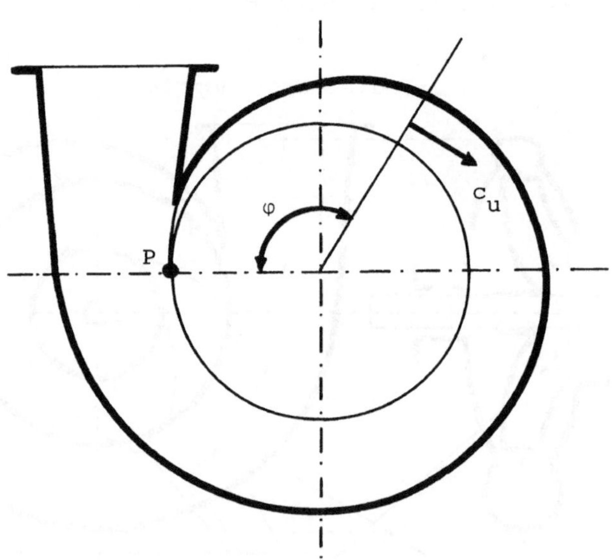

Fig. 2 Volute

A point P, called the volute tongue, marks the begin of the volute. Its distance r_0 from the impeller axis is assumed to be known. The volute width $2x_0$ is constant along the angle φ and is known. The cross section S_φ of the volute at some angle φ is symmetric about the r-axis and its boundary is described by a function $x = x(r)$.

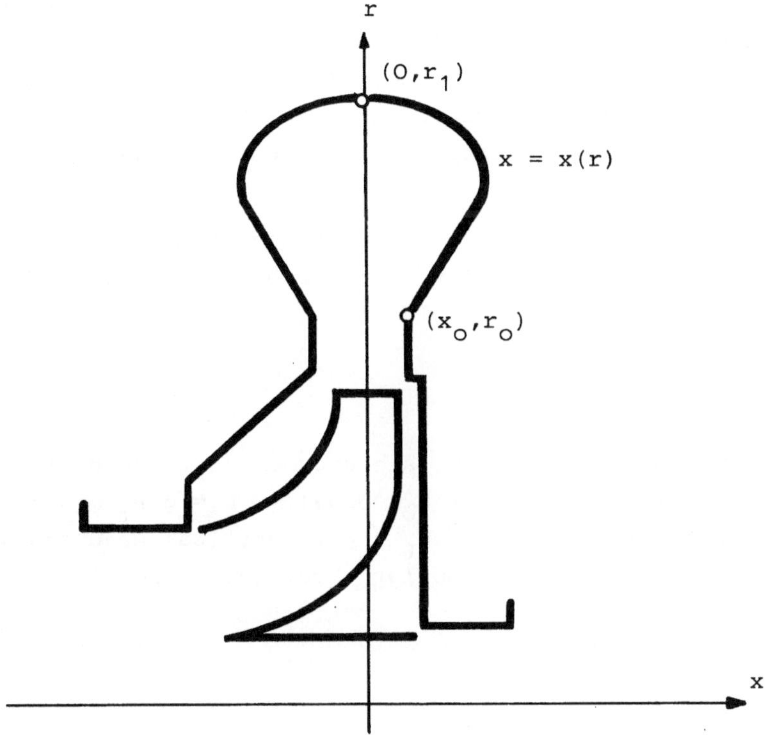

Fig. 3   Cross section S_φ

Let Q_φ be the specific flow rate of liquid (or flux) across the cross section S_φ. It is easy to see that

$$(2.1) \qquad Q_\varphi = \iint_{S_\varphi} c_u \, dxdr$$

with c_u the component of the velocity normal to S_φ. It is assumed that liquid leaves the impeller uniformly along φ. That implies a linearly increasing flux:

(2.2) $\quad Q_\varphi = \dfrac{\varphi}{360^\circ} \cdot Q$

with Q the flux leaving the volute at angle $\varphi = 360^\circ$ (Q is called the capacity of the pump.)

For a pump of high efficiency the losses of energy in the pump are to be low. The following three principles are supposed to guarantee small losses. (The first two principles postulate a uniform flow of liquid in the volute. Since any change causes losses, uniformity is believed to keep the losses low. The third principle deals with friction losses. For details, see [4], [5], [7].)

a. The principle of constant angular momentum, after Pfleiderer [4]: The specific absolute angular momentum D is assumed to be constant within the volute:

(2.3) $\quad D = c_u \cdot r$

with c_u the normal speed (see above) and r the distance from the x-axis at a point of S_φ. Usually, $D = g\, H_t/\omega$ with g the acceleration of gravity, H_t the theoretical head and ω the angular speed of the impeller. From (2.1) and (2.3), we have

(2.4) $\quad Q_\varphi = D \cdot b_\varphi$

with

$$b_\varphi = \iint_{S_\varphi} \frac{1}{r}\, dx\, dr = 2 \int_{r_o}^{r_1} \frac{1}{r}\, x(r)\, dr$$

b. The principle of constant average speed, after Stepanoff [5]. It is assumed that the average speed

(2.5) $\quad c = \dfrac{1}{F_\varphi} \cdot \iint_{S_\varphi} c_u\, dx\, dr$

is constant in φ, where F_φ is the area of S_φ, i.e.:

$$F_\varphi = 2 \int_{r_o}^{r_1} x(r)\,dr.$$

The average speed is determined by the formula $c = K\sqrt{2gH}$ with H the pump head and K an experimental design factor. From (2.1) and (2.5) we have

$$(2.6) \quad Q_\varphi = F_\varphi \cdot c$$

c. The hydraulic radius is maximized. Let U_φ be the wetted perimeter of S_φ, i.e.

$$U_\varphi = 2 \int_{r_o}^{r_1} \sqrt{1 + \dot{x}(r)^2}\,dr.$$

The hydraulic radius $2F_\varphi/U_\varphi$ is a rough measure for the friction losses within the volute. In order to minimize these losses, the cross section S_φ is required to have a hydraulic radius as large as possible:

$$\frac{2F_\varphi}{U_\varphi} \to \max.$$

For a fixed angle φ, values for b_φ and F_φ are given by

$$b_\varphi = \frac{Q}{D}\frac{\varphi}{360^o} \quad \text{and} \quad F_\varphi = \frac{Q}{c}\frac{\varphi}{360^o}$$

according to (2.2), (2.4), (2.6).

3   The mathematical model

The assumptions on the design of the volute lead to the following problem in the calculus of variation:

Given data: r_o, x_o, F_φ, b_φ (φ is fixed)

Find the end point r_1 and a function $x(r)$ on $[r_o, r_1]$ such that

$$(3.1) \quad \frac{1}{F_\varphi} \int_{r_o}^{r_1} \sqrt{1+\dot{x}(r)^2} \; dr \to \min$$

subject to the linear constraints

$$(3.2) \quad 2 \int_{r_o}^{r_1} x(r) dr = F_\varphi,$$

$$(3.3) \quad 2 \int_{r_o}^{r_1} \frac{1}{r} x(r) dr = b_\varphi,$$

and conditions for the fixed end point r_o and the variable end point r_1

$$(3.4) \quad x(r_o) = x_o,$$

$$(3.5) \quad x(r_1) = 0.$$

Necessary conditions for this problem with variable end point are (see [2])

a. the Euler-Lagrange equation:

$$(3.6) \quad \frac{\partial f}{\partial x} (x,\dot{x},r) - \frac{d}{dr} \frac{\partial f}{\partial \dot{x}} (x,\dot{x},r) = 0$$

with $f(x,\dot{x},r) = \frac{1}{F} \sqrt{1+\dot{x}^2} + \lambda_1 x + \lambda_2 \frac{1}{r} x$, λ_1 and λ_2 are multipliers associated to the constraints.

b. the transversality condition:

$$(3.7) \quad \left[f(x,\dot{x},r) - \dot{x} \frac{\partial f}{\partial \dot{x}} (x,\dot{x},r) \right]\Big|_{r=r_1} = 0$$

Equation (3.6) leads to the differential equation

$$(3.8) \quad \frac{d}{dr} \frac{\dot{x}}{\sqrt{1+\dot{x}^2}} = \mu_1 + \frac{\mu_2}{r}$$

with $\mu_1 = \lambda_1 F$ and $\mu_2 = \lambda_2 F$. Condition (3.7) is satisfied only if

$$(3.9) \quad \dot{x}(r_1) = -\infty.$$

The solution of (3.8), (3.9) with (3.5) in terms of the parameters μ_1, μ_2 and r_1 is

$$(3.10) \quad x(r) = \int_{r_1}^{r} \frac{\psi(r)}{\sqrt{1-\psi(r)^2}} \, dr$$

where $\psi(r) = \mu_1(r-r_1) + \mu_2 \ln \frac{r}{r_1} - 1$. By using the above representation of $x(r)$, the linear constraints (3.2), (3.3) and the remaining condition at the fixed end point (3.4) lead to the following three equations in the three unknowns μ_1, μ_2 and r_1:

$$2 \int_{r_o}^{r_1} (r-r_o) \frac{\psi(r)}{\sqrt{1-\psi(r)^2}} \, dr + F_\varphi = 0$$

$$(3.11) \quad 2 \int_{r_o}^{r_1} \ln \frac{r}{r_o} \cdot \frac{\psi(r)}{\sqrt{1-\psi(r)^2}} \, dr + b_\varphi = 0$$

$$\int_{r_o}^{r_1} \frac{\psi(r)}{\sqrt{1-\psi(r)^2}} \, dr + x_o = 0$$

These equations are obtained by integration by parts.

$\underline{\text{Remark}}$: It is believed that the equations (3.11) are also sufficient for a solution to the problem of the calculus of variation. This has not yet been analyzed by the author of this paper.

4 The numerical method

The system (3.11) of three equations, shortly denoted by

$$G(z) = 0 \quad \text{with } z = (\mu_1, \mu_2, r_1),$$

is solved by Newton's method:

$$(4.1) \quad G'(z^i)\Delta z^i = -G(z^i), \quad z^{i+1} = z^i + \Delta z^i$$

with $G'(z)$ the Jacobian of G at the point z.

Evaluation of G(z) and G'(z)

For each iteration, $G(z^i)$ and $G'(z^i)$ have to be evaluated. This leads to the numerical approximation of integrals of the form

$$\int_0^1 f(s)\,ds \quad \text{with } f(s) = \frac{1}{\sqrt{s}}\, g(\sqrt{s}) \text{ and } g \text{ bounded.}$$

(if normalized to the interval [0,1] by the transformation $r = r_1(1-s) + r_0 s$). The integrand has a singularity at $s = 0$. By setting $s = t^2$ this singularity vanishes:

$$\int_0^1 f(s)\,ds = 2 \int_0^1 g(t)\,dt.$$

The remaining integral is approximated by Rombergs' method, see [6].

The initial value for Newton's method

A rather good initial value z^O is necessary for (4.1) to converge. This is partly due to the form of the denominator occurring in (3.1o), which makes it necessary to restrict the domain for z to valueswith $-1 \leq \psi(r) \leq 1$.

A general technique for determining an appropriate initial value z^O is the homotopy method (see [3]). The original problem is imbedded into a family of problems by introducing an additional homotopy parameter. For this problem, the angle φ is chosen as the homotopy parameter. For one particular value of φ, say φ^O, the solution of the problem is assumed to be known. This solution is an appropriate initial guess for the problem with a slightly changed angle $\varphi^O + \Delta\varphi$. Newton's method gives the solution for this new angle. By repeating this procedure one eventually obtains a solution for any angle φ.

It remains to find a starting angle φ^O for which the solution is known. For this, the problem in the calculus of variation of Section 3 without the second linear constraint (3.3) is considered. It can be solved explicitly for arbitrary angle φ. The solutions $x(r)$ are circles. It is easy to determine that angle φ^O for which one of these circles also satisfies the remaining constraint (3.3). This angle is chosen as starting point of the homotopy method mentioned above.

Once the parameters $z = (\mu_1, \mu_2, r_1)$ are known, the function $x(r)$ is determined at an arbitrary point r by evaluating the integral in (3.1o) by Romberg's method.

5   **Discussion of the numerical results and a revised model**

The model of Section 3 was tested with the following data

$$(5.1) \quad \frac{Q}{C} = 1928,2 \text{ cm}^2, \quad \frac{Q}{D} = 12,6 \text{ cm}^2, \quad x_O = 8 \text{ cm}, \quad r_O = 133 \text{ cm}$$

It turned out that for small angles φ completely unrealistic solutions $x(r)$ occurred, caused by the fact that the two constraints (3.2) and (3.3) are not compatible for small angles. Therefore, the model has to be changed. It is proposed to drop the constraint

$$2 \int_{r_o}^{r_1} x(r) = F_\varphi$$

for small angles. A similar analysis as before leads to the following set of equations for the new model (we use the same notation as in Section 3):

$$\int_{r_o}^{r_1} \frac{1}{\sqrt{1-\psi(r)^2}}\, dr - \mu_1 \int_{r_o}^{r_1} (r-r_o)\, \frac{\psi(r)}{\sqrt{1-\psi(r)^2}}\, dr = 0$$

$$(5.2) \quad 2 \int_{r_o}^{r_1} \ln \frac{r}{r_o}\, \frac{\psi(r)}{\sqrt{1-\psi(r)^2}}\, dr + b_\varphi = 0$$

$$\int_{r_o}^{r_1} \frac{\psi(r)}{\sqrt{1-\psi(r)^2}}\, dr + x_o = 0$$

These equations are obtained by regarding F_φ as an additional unknown occuring in the objective (3.1) and in the constraint (3.2).

As before, the system (5.2) is solved by Newton's method. Fig. 5 shows the obtained cross sections, Fig. 4 the shape of the volute for the data (5.1). The boundary of the cross section is a circle for some angle φ^o between 270° and 300°. For angles larger than or equal to 270°, the equations (3.11) of the original model were used to determine $x(r)$. This leads to the last four cross sections shown in Fig. 5. For smaller angles the computation was continued by solving the equations (5.2) of the revised model. This change to the revised model was chosen so early in order to guarantee a smooth transition. Fig. 6 confirms this by showing that the condition (3.2) is satisfied

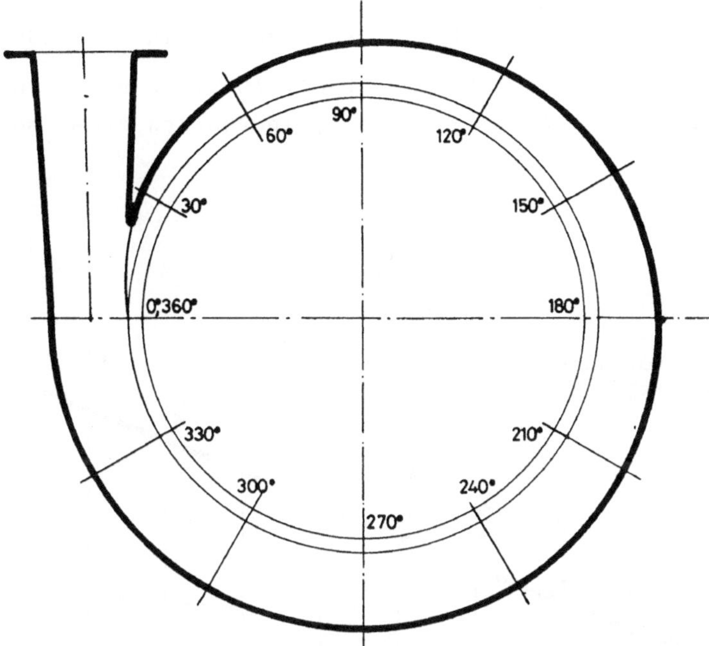

Fig. 4 The calculated shape of the volute

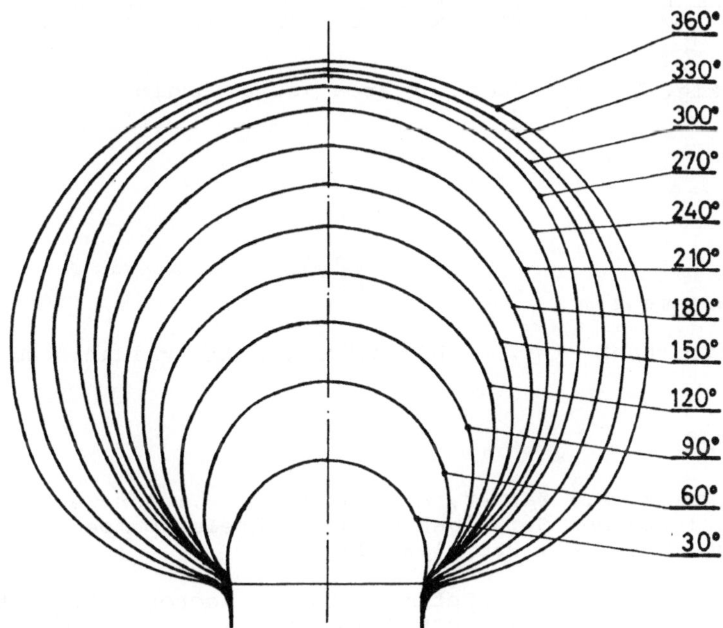

Fig. 5 The calculated cross sections

to a high degree although it is not contained in the model as a constraint for angles φ less than 270°. There is almost no deviation from the case of a linearly increasing cross section area F_φ, as required in the original model.

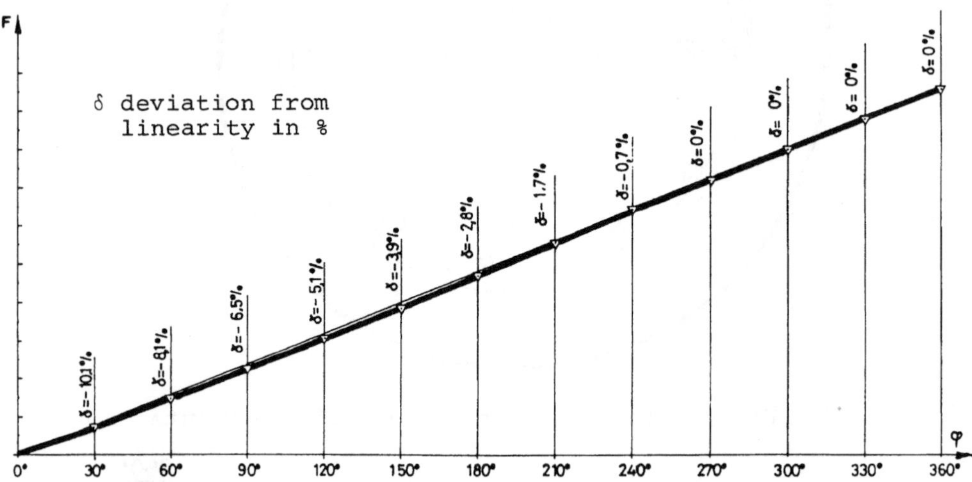

Fig. 6   cross section area versus angle

Acknowledgement

We wish to thank Prof.Dr. Hj.Wacker for the valuable advice.

References

[1]   Kowalik, M.; Zarzer, E.; Zulehner, W.: Zur Gestaltung eines
      verlustarmen Spiralgehäuses für Kreiselpumpen. Forsch.
      Ing.-Wes. 46 (1980) 196-199

[2]   Luenberger, D.G.: Optimization by vector space methods.
      New York: John Wiley & Sons 1969

[3]   Ortega, J.M.; Rheinbold, W.C.: Iterative solution of non-
      linear equations in several variables. New York: Academic
      Press 1970

[4]   Pfleiderer, C.: Die Kreiselpumpe für Flüssigkeiten und
Gase. 5. Aufl. Berlin-Göttingen-Heidelberg: Springer 1961

[5]   Stepanoff, A.J.: Centrifugal and axial flow pumps. 2nd ed.
New York: John Wiley & Sons 1957

[6]   Stoer, J.; Bulirsch, R.: Introduction to numerical analysis.
New York: Springer 1980

[7]   Troskolanski, A.T.; Lazarkiewicz, S.: Kreiselpumpen, Berechnung und Konstruktion. Basel-Stuttgart: Birkhäuser
1976

NUMERICAL CALCULATION OF SEPARATION PROCESSES

D.Auzinger, L.Peer, Hj.Wacker, W.Zulehner

1   Introduction

Distillation is one of the most common separation processes carried out in industry. Separation by distillation is
based on the different volatilities of the components of a mixture. If a liquid mixture of two or more components is partially
vaporized or if a vapor mixture is partially condensed, then, in
either case, the vapor product is richer in the component of
highest volatility while the liquid product is richer in the
component of lowest volatility. A one-stage vaporisation or
condensation process gives products of a certain limited purity.
In order to increase product purity, a sequence of successive
vaporizations of the liquid products, resp. a sequence of successive condensations of the vapor products, are used. An eco-
nomic way for carrying out such a multistage distillation process
is a distillation column, see Fig. 1. The column consists of a
number of plates, on each of which vapor and liquid are in
equilibrium with each other. The mixture enters the column at
some plate (feed). Under the force of gravity the liquid flows
downwards. A heater, called the reboiler, at the bottom of the
column vaporizes a portion of the liquid. The resulting vapor
flows upwards. A condensor at the top of the column liquifies a
portion of vapor which is returned to the column (reflux). The
remaining portion of the overhead vapor is called the distillate.

A typical problem specification for a distillation column:
Given data: design of the column (number of plates, location of
feed), pressure within the column, feed (flow rate, enthalpy,
composition), distillate (flow rate), reflux (flow rate).
Problem: Determine the composition of the distillate.

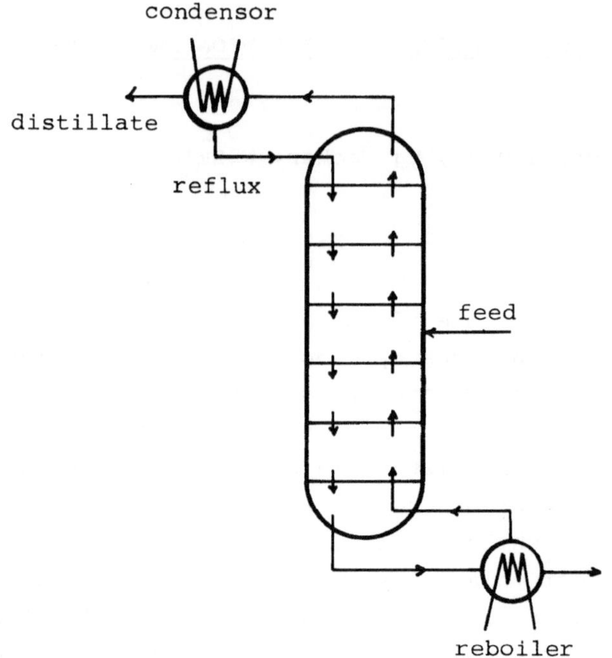

Fig. 1. Distillation column

For further details on distillation and other separation
processes, see [3].

In the following section, first the governing equations
of a slightly more general distillation column are described.
Then the model is extended by considering columns with so-called
pumparounds and side strippers. The resulting system of nonlinear
equations is studied in Section 3. In particular, the sparsity
pattern of the Jacobian is analyzed. Section 4 contains the
description of the numerical methods. Another separation process,
namely absorption, is shortly studied in Section 5. More general
separation processes are discussed in Section 6. Industrial
plants for these processes consist of a number of interlinked
units such as distillation columns. A numerical approach is

presented for these problems. An example, the production of
oxygen from air, concludes the paper.

There has been a long term cooperation on the subject of
this paper with the local branch of the steel company VOEST-
ALPINE AG, in particular, with F.Kokert.

2 The physical model of a distillation column

2.1 The basic model

A slightly more general distillation column is considered:
Streams can be fed to the column at different points (multiple
feed), liquid and vapor streams of given quantities can be with-
drawn from any stage (sidestreams) and, finally, given amounts
of heat can be added to or subtracted from any stage.

Distillation is considered as an equilibrium-stage process,
i.e. the liquid and vapor phases leaving any stage are in equilibrium with each other. Pressure is assumed to be given at each
stage. Then the equilibrium is described by the liquid and vapor
flow rates of each component and the temperature at each stage.

Notation. Let N be the total number of stages including
the reboiler (first stage) and the condensor (last stage), see
Fig. 2, and let M be the number of components involved in the
distillation process. Then, for $n = 1,2,\ldots,N$ and $j = 1,2,\ldots,M$,
the following unknowns describing the state of equilibrium are
introduced:

l_{nj} the liquid flow rate of component j from stage n to stage
n-1 (mol/time)

v_{nj} the vapor flow rate of component j from stage n to stage
n+1 (mol/time)

T_n the temperature of stage n

In addition, the total liquid and vapor flow rates are denoted
by L_n resp. V_n, stagewise:

$$L_n = \sum_{j=1}^{M} l_{nj}, \qquad V_n = \sum_{j=1}^{M} v_{nj}.$$

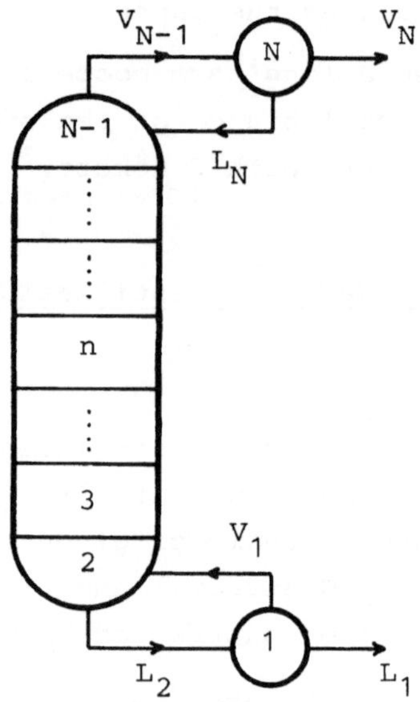

Fig. 2

The basic model for the distillation column consists of three sets of equations:

Component mass balances

Mass input must be equal to mass output for each component at each stage, see Fig. 3.

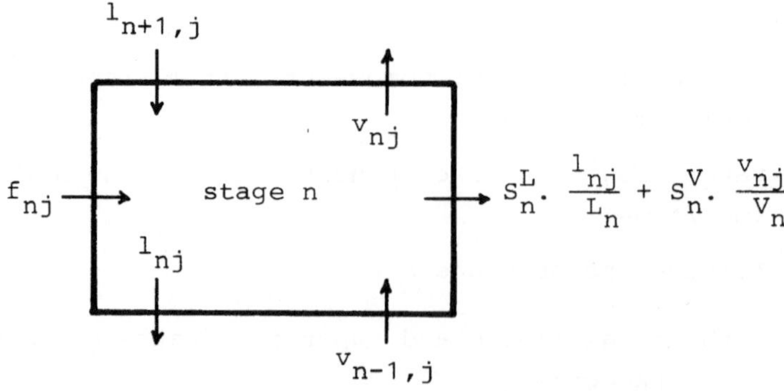

Fig. 3. Mass balances

Therefore, we have

$$b_{nj} := (1+\frac{S_n^L}{L_n})l_{nj} + (1+\frac{S_n^V}{V_n})v_{nj} - l_{n+1,j} - v_{n-1,j} - f_{n,j} = 0$$

with given f_{nj} the feed flow rate of component j at stage n

$\quad S_n^L \quad$ the total sidestream liquid flow rate at stage n

$\quad S_n^V \quad$ the total sidestream vapor flow rate at stage n

Enthalpy balances

Energy input must be equal to energy output at each stage, see Fig. 4.

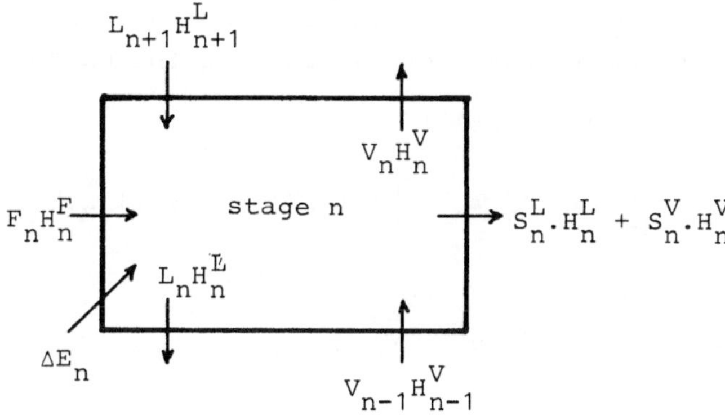

Fig. 4. Enthalpy balances

Therefore, we have

$$E_n := (L_n + S_n^L)H_n^L + (V_n + S_n^V)H_n^V - L_{n+1}H_{n+1}^L - V_{n-1}H_{n-1}^V - F_n H_n^F - \Delta E_n = 0$$

with $\quad H_n^L \quad$ the molar enthalpy of liquid transported from stage n to stage n-1

$\quad H_n^V \quad$ the molar enthalpy of vapor transported from stage n to stage n+1

$\quad F_n \quad$ the given total feed flow rate at stage n $(F_n = \sum_{j=1}^{M} f_{nj})$

H_n^F the given molar enthalpy of feed at stage n

ΔE_n given net amount of heat added to stage n per time.

Vapor-liquid equilibrium relationship

The vapor and liquid phase of component j are in equilibrium with each other. This leads to the condition

$$g_{nj} : = k_{nj} \cdot \frac{l_{nj}}{L_n} - \frac{v_{nj}}{V_n} = 0$$

with k_{nj} the equilibrium ratio.

H_n^L, H_n^V and k_{nj} are given functions of (known pressure,) temperature and the composition of the liquid and vapor phase. In this paper their particular form is based on the UNIQUAC model, see e.g. [1]:

$$H_n^L = \sum_{j=1}^{M} (a_j + b_j \cdot T_n) \, \frac{l_{nj}}{L_n} + H^E(l_{n1}, \ldots, l_{nM}, T_n)$$

$$H_n^V = \sum_{j=1}^{M} (c_j + d_j T_n) \cdot \frac{v_{nj}}{V_n}$$

$$k_{nj} = \frac{P_j^O(T_n)}{P_n} \cdot \frac{\gamma_j(l_{n1}, \ldots, l_{nM}, T_n)}{\varphi_j(v_{n1}, \ldots, v_{nM}, T_n)}$$

with H^E the molar excess enthalpy

       P_j^O the vapor pressure of pure component j

       P_n pressure at stage n

       γ_j liquid-phase activity coefficient of component j

       φ_j vapor-phase fugacity coefficient of component j

Instead of enthalpy balances for stage 1 and N with unknown heat transfer rates ΔE_1 and ΔE_N, we use the equation

$$E_N: = L_N - r.D = 0$$

with given

    D the flow rate of distillate ($V_N = D$)

    r the reflux ratio

and, for reasons that become clear later, not $V_N = D$ but the overall mass balance

$$E_1: = L_1 + D + \sum_{n=1}^{N} (S_n^V + S_n^L - F_n) = 0.$$

Summary: The basic model for a distillation column consists of the $(2M+1)N$ given equations

$$b_{nj} = 0, \quad E_n = 0, \quad g_{nj} = 0 \quad \text{for } n = 1,\ldots,N, \; j = 1,\ldots,M$$

in $(2M+1)N$ unknowns

$$l_{nj}, \; v_{nj}, \; T_n \qquad\qquad \text{for } n = 1,\ldots,N, \; j = 1,\ldots,M$$

2.2   Pumparounds

The basic model is now extended by considering pumparounds, see Fig. 5.

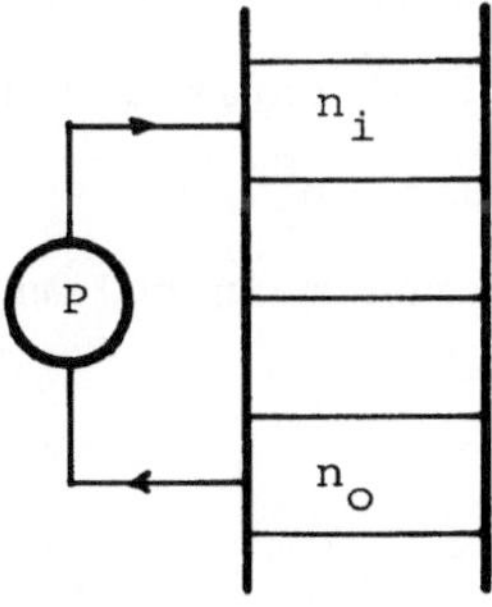

Fig. 5. Pumparound

A pumparound removes a liquid stream of given quantity L^P from plate n_o and, after cooling or heating, returns it at some plate n_i. This changes the component mass balances resp. enthalpy balances of the involved plates:

Component mass balances

$$b_{nj} \Leftarrow b_{nj} + L^P \frac{l_{n_o,j}}{L_{n_o}} \qquad \text{for } n = n_o$$

$$b_{nj} \Leftarrow b_{nj} - L^P \frac{l_{n_o,j}}{L_{n_o}} \qquad \text{for } n = n_i$$

Enthalpy balances

$$E_n \Leftarrow E_n + L^P H^L_{n_o} \qquad \text{for } n = n_o$$

$$E_n \Leftarrow E_n - L^P H^L_{n_o} \qquad \text{for } n = n_i$$

(Read A $\Leftarrow$ B as "A is replaced by B".)

2.3   <u>Side stripper</u>

A further extension is made by considering certain types of side strippers, see Fig. 6.
Liquid and vapor streams of given quantities L^S and V^S, removed from the plates n_o^L resp. n_o^V enter a side stripper at the top plate n_t^S resp. bottom plate n_b^S. The top product of the side stripper is returned to the main column at plate n_i.

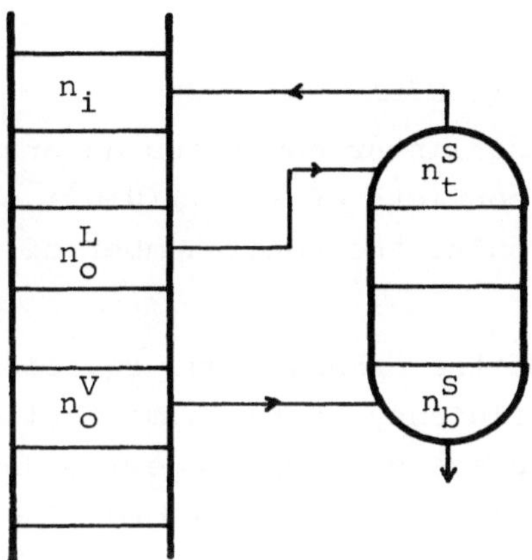

Fig. 6. Side stripper

The model for the side stripper only is completely analogous to the basic model for the main column. The linkage between the two columns leads to the following changes:

Component mass balances:

$$b_{nj} \; \Leftarrow \; b_{nj} + L^S \cdot \frac{l_{n_o^L,j}}{L_{n_o^L}} \qquad \text{for } n = n_o^L$$

$$b_{nj} \; \Leftarrow \; b_{nj} + V^S \cdot \frac{v_{n_o^V,j}}{V_{n_o^V}} \qquad \text{for } n = n_o^V$$

$$b_{nj} \; \Leftarrow \; b_{nj} - v_{n_t^S,j} \qquad \text{for } n = n_i$$

$$b_{nj} \; \Leftarrow \; b_{nj} - L^S \cdot \frac{l_{n_o^L,j}}{L_{n_o^L}} \qquad \text{for } n = n_t^S$$

$$b_{nj} \; \Leftarrow \; b_{nj} - V^S \cdot \frac{v_{n_o^V,j}}{V_{n_o^V}} \qquad \text{for } n = n_b^S$$

Enthalpy balances are analogously modified at the same plates.

3   The mathematical model

The system of nonlinear equations describing the general problem of Section 2 consists of $(2M+1)(N+N^S)$ equations and unknowns, where N^S denotes the total number of plates occuring in side strippers.

The ordering of the unknowns and equations is of great importance for the efficiency of the numerical method. If the unknowns and equations are grouped stagewise and if these blocks are assembled according to the natural ordering of the stages, see Fig. 2, then the Jacobian for the basic model is block tridiagonal, because only up to three successive stages are involved for each equation. (The block tridiagonal structure would have been destroyed if the equation $V_N = D$ had been used instead of the overall mass balance $E_1 = 0$.)

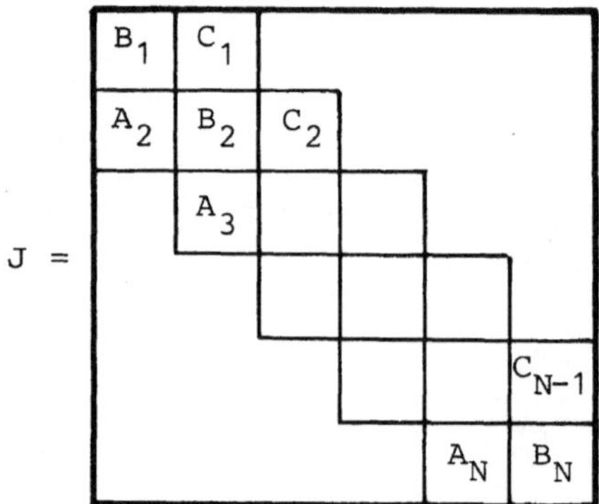

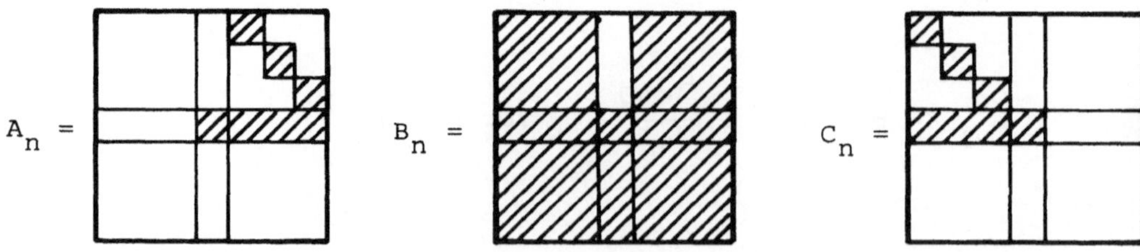

Fig. 7. Structure of the Jacobian for the main column. The shaded regions mark the nonzero elements.

The upper and lower bandwidth of this matrix is, in general,
3M+1. By ordering the unknowns, resp. the equations, of stage n
in the way $(l_{n1}, \ldots, l_{nM}, T_n, v_{n1}, \ldots, v_{nM})$, resp.
$(b_{n1}, \ldots, b_{nM}, E_n, g_{n1}, \ldots, g_{nM}) = 0$, the bandwidth is reduced to
2M+1. Fig. 7 shows the structure of the N-by-N block tridiagonal
Jacobian with (2M+1)-by-(2M+1) submatrices, which are further
partitioned into M,1 and M rows, resp. columns.
Each pumparound leads to an additional nonzero block of the
Jacobian, see Fig. 8.

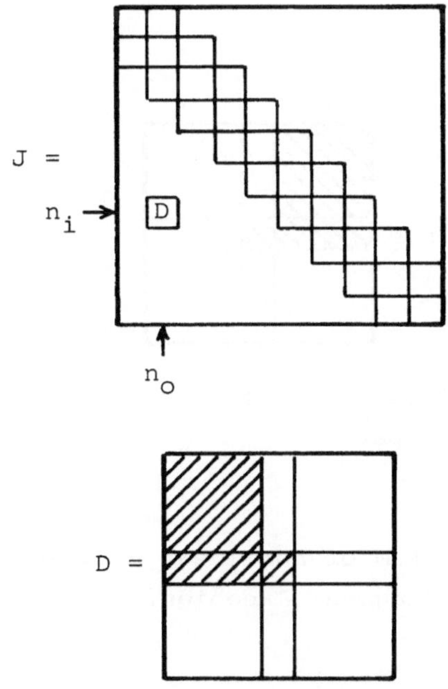

Fig. 8. Structure of the Jacobian for
a main column with one pumparound

For each side stripper, a block tridiagonal matrix is augmented
to the Jacobian of the main column and three additional blocks
occur, see Fig. 9.

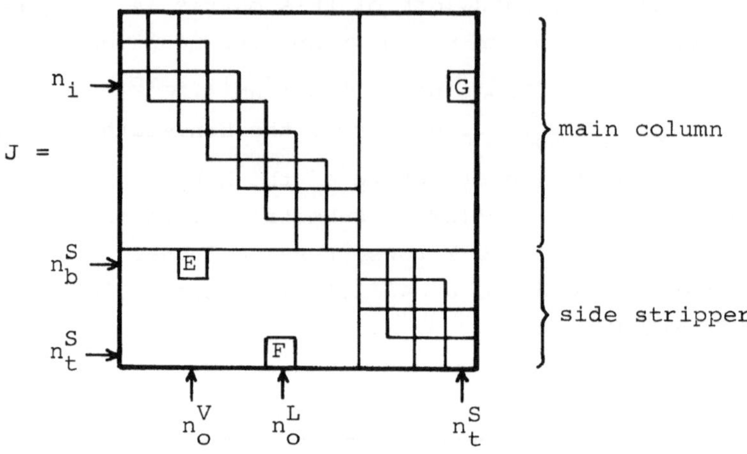

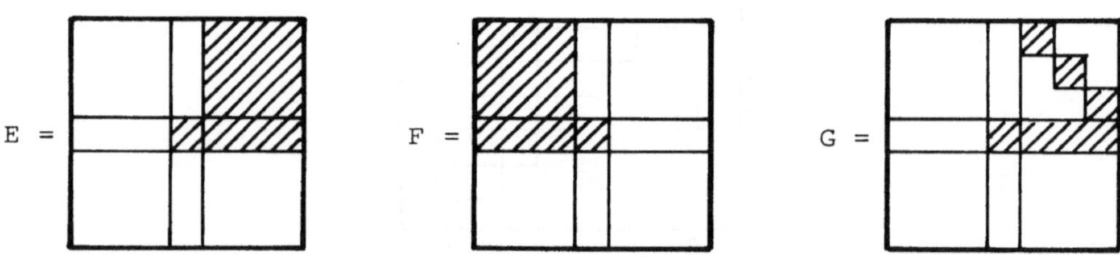

Fig. 9. Structure of the Jacobian for a main column with
one side stripper

In summary, a sparse system of nonlinear equations is obtained
with a mainly block tridiagonal Jacobian.

4   Numerical methods

The system of nonlinear equations of Section 2, shortly
denoted by

$$f(x) = 0$$

is solved by replacing it by a sequence of linear problems

$$J^k \cdot (x^{k+1} - x^k) + f(x^k) = 0, \quad k = 0,1,\ldots$$

In particular, we discuss two Newton-like methods. Direct as well as iterative procedures for solving the occuring linear systems are studied. Strategies for improving convergence complete the section.

Nonlinear solvers

For Newton's method, J^k coincides with the Jacobian $f'(x^k)$. The complexity of the underlying UNIQUAC model does not allow an efficient explicit evaluation of the Jacobian. To overcome these difficulties, the matrix is split into two parts, see Westman et.al. [9]:

$$f'(x^k) = C(x^k) + A(x^k).$$

$C(x^k)$ contains the easy-to-calculate derivatives and can be evaluated explicitly. $A(x^k)$ consists of all terms of the Jacobian, where derivatives of H^E, γ_j or φ_j occur. For the matrix J^k, we set

$$J^k = C(x^k) + A^k$$

with some approximation A^k for $A(x^k)$. Two alternatives are considered for A^k: $A(x^k)$ is approximated either

a. by the Schubert update, see [6], which is an adaptation of Broyden's rank-1 method to sparsity or

b. by using finite differences for the derivatives of H^E, γ_j and φ_j. (This finite difference approximation requires considerably less numerical work than a direct finite difference approximation of $f'(x^k)$.)

The methods based on these approximations are called the hybrid quasi-Newton method (hQNM) and the hybrid finite difference Newton method (hFDNM), respectively.

Comparison of the two methods:

The numerical effort for one iterate of hQNM is dominated by the amount of time t_1 for solving the linear system. hFDNM requires some additional time t_e for evaluating the approximate Jacobian. (Notice, that the numerical effort for solving the linear system is the same for hQNM and hFDNM, because hQNM does not allow an efficient direct update of the L-U factorization.) Numerical tests showed that the ratio $t_1 : t_e$ is about 1 : 3 for the case of a main column only (basic model with UNIQUAC). Hence, about four times more iterates can be performed with hQNM than with hFDNM within the same time, which usually outweighs the lower rate of convergence for hQNM. Therefore, Westman et.al. [9] recommended hQNM in this case. However, the numerical effort for solving the linear system increases considerably for main columns with pumparounds and side strippers. The ratio $t_1 : t_e$ quickly changes from 1 : 3 to 1 : 1 for systems of moderate complexity, even to 3 : 1 for complex configurations. Therefore, we advocate for hFDNM in these cases.

Linear solvers

Either nonlinear solver generates a linear algebraic system at each iteration step.

A direct solver. In the absence of pumparounds and side strippers two approaches are usually taken to solve the linear system. One way is to stress the block tridiagonal structure of the matrix and use a block Thomas algorithm. For the other approach the matrix is considered as a band matrix and a band Gaussian L-U factorization is used. A general code for solving the linear system in the first way requires about $18\frac{2}{3}$ NM^3 operations. The number of operations reduces to 8 NM^3 (resp. to 16 NM^3 with partial pivoting) for the second approach.

As an alternative, we propose a Gaussian L-U factorization with incomplete partial pivoting (shortly: ippLU) under exploitation of the special structure of the matrix. Roughly speaking, partial pivoting is used, whenever it does not essentially destroy the sparsity pattern. In particular, the diagonal elements in the right upper part of A_n, see Fig. 7, are chosen for pivoting. This method is reasonably stable and requires about $5\frac{5}{6}$ NM^3 operations for problems with a main column only. It naturally extends to a method for the full problem of Section 2, where off-tridiagonal blocks lead to a fill-up of certain rows and columns.

Another solver for almost block tridiagonal systems was suggested by Kubiček et.al. in [4]: The original equation is transformed into a set of equations with a block tridiagonal structure. If we assume that these block tridiagonal systems are solved by ippLU, then, for the example of a main column with one pumparound, the number of operations for Kubiček's method ranges from $1o\frac{5}{6}$ NM^3 to $13\frac{5}{6}$ NM^3 depending on the position of the pump-around. The proposed extended ippLU requires between $5\frac{5}{6}$ NM^3 and $1o\frac{1}{3}$ NM^3 operations for this example. The superiority of ippLU to Kubiček's method is even higher for more complex problems.

An iterative solver. A very natural splitting of the matrix J^k is suggested by the special structure of the problem:

$$J^k = T^k + R^k$$

with T^k is block tridiagonal and R^k is extremely sparse. This leads to the iterative method SPLIT

$$T^k.d^{k,l+1} + R^k.d^{k,l} + f(x^k) = 0, \quad l = 0,1,\ldots$$

for determining $d^k = x^{k+1}-x^k$. Each step requires the solution of a block tridiagonal system, which is done by ippLU.

Comparison of direct and iterative solver

In general, the iterative solver SPLIT needs considerably less computer memory than the direct solver ippLU. The computational effort for the direct solver increases with the number of off-band blocks and their distance from the diagonal. Numerical tests show that the rate of convergence of the iterative solver SPLIT mainly depends on the magnitude of the off-band blocks but not on their position. The following example was chosen to demonstrate the influence of the magnitude of the off-band blocks on the rate of convergence.

Example: (for demonstrational purposes only)
Separation of methanol-water by a distillation column (main column: 3o plates, one pumparound from plate 2 to plate 28, see Fig. 1o)

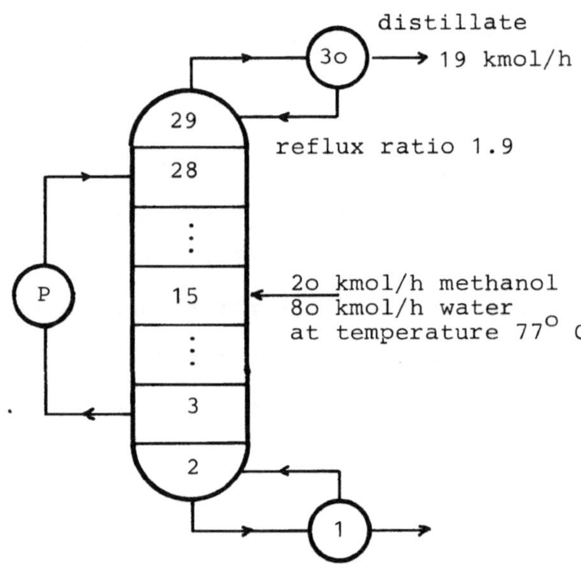

Fig. 1o. Distillation of methanol-water
(constant pressure 1 bar)

Compared to ippLU, the iterative solver saves about 5 percent of total cpu-time if L^p = 12 kmol/h (L^p flow rate of pumped stream). For L^p = 3o kmol/h, the direct solver is faster by about 8 percent.

The described behaviour was typical for a series of further test problems, so we may say, that the iterative solver is recommended for systems with blocks far off a dominant block tridiagonal.

Improvements

Initial value calculations. Of great importance for the efficiency of either method (hQNM or hFDNM) is a good initial approximation of the solution. Using estimates of the temperature of top, bottom and feed stages, initial values for T_n are calculated by linear interpolation. A plate to plate overall mass balance calculation gives approximations to L_n and V_n. These total flow rates are partitioned into component flow rates l_{nj} and v_{nj} by means of an idealized vapor-liquid relationship.

Damping. An inexact line search is incorporated in hQNM and hFDNM in order to increase the region of convergence.

Non-negativity conditions. There are natural lower bounds for the variables

$$l_{nj} \geq 0, \quad v_{nj} \geq 0, \quad T \geq 0 \quad \text{(in Kelvin)}$$

Whenever an iterate x^{k+1} violates these conditions, negative components are replaced by zero before continuing the method. This is particularly important to prevent the evaluation of H^E, γ_j and φ_j at meaningless points which could completely destroy convergence.

Homotopy methods. Even an improved initial value calculation does not necessarily ensure convergence. In this case homotopy methods may help to overcome the difficulties: The basic idea is to continuously deform the given problem into a simpler one. Then the required solution of the given problem is obtained by following a continuous path starting at a solution of the simpler problem. For details, see e.g. [8], [2]. The deformation can be done, for example, by changing some internal parameters of the problem. In particular the reflux ratio, the heat transfer ΔE_n, flow rates of side streams, pumped streams or exchange streams between main column and side stripper are used.

5   Absorber

Another important multistage separation device is an
absorber, see Fig. 11.

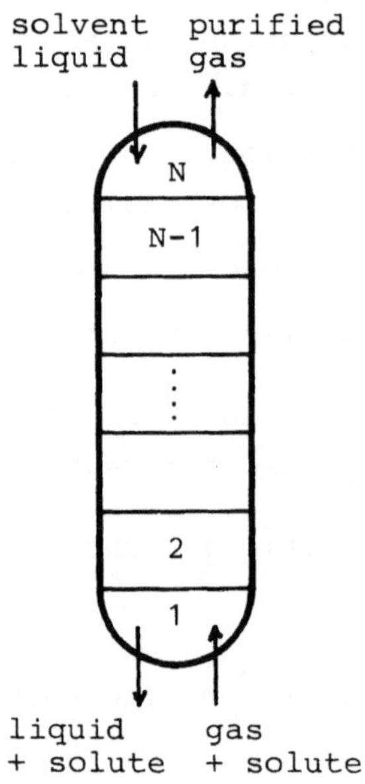

Fig. 11. Absorber

The separation is based on the ability of a solvent liquid
to absorb a soluble impurity (solute) from a gas. Countercurrent
streams are obtained by a liquid feed at the top and a gas feed
at the bottom of the column. The process is described by the
same equations as for a distillation column, with two exceptions:
The enthalpy balances are used for stage 1 and N instead of the
reflux condition and the overall mass balance.

As before, extensions of the basic model by multiple
feeds, sidestreams, added heat transfer, pumparounds and side
strippers are considered. The numerical treatment of the resulting system of equations is analogous to the case of distillation
columns.

6   <u>General separation processes</u>

Industrial plants for carrying out more general separation processes may consist of a series of interlinked distillation columns, absorbers and other devices, such as:

a. Counter-flow heat exchangers (Liquid and/or vapor streams of different temperatures flow countercurrently through parallel tubes along which heat is exchanged. For given input temperatures, the calculation of the output temperatures leads to the solution of a boundary value problem for a system of ordinary differential equations whose right hand side may have jumps.)

b. Simple units: mixers (streams are mixed), splitters (a stream is split into two or more separate streams), compressors (pressure is increased), turbines (pressure is decreased), heaters and coolers. For these simple devices, the calculation of the output is straight forward for given input.

The complicated structure of such a plant is best described by introducing a directed graph (flowsheet) whose nodes are the various devices and whose edges represent the material stream connections between these devices. It is assumed that, for each single device of the plant, a working program exists for computing the output streams for given input streams. In order to compute the overall solution of the plant, we proceed in two steps:

a. Partitioning: The plant is partitioned into so-called irreducible subsystems which are sequentially coupled. This weak form of coupling allows to compute the overall solution by sequentially solving the subsystems separately. The irreducible subsystems are easily determined either by inspection or, for more complicated problems, by means of the elementary circuits of the graph. (An elementary circuit is a closed path which contains no node twice. For an algorithm, see [7].) The crucial part is

b. The solution of the irreducible subsystems. An irreducible system consisting of two or more units cannot be calculated sequentially, because it contains at least one elementary circuit. Fig. 12 shows the simplest case of such a system.

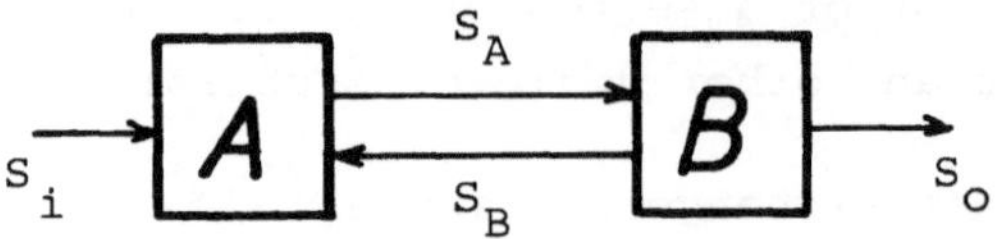

Fig. 12

The two devices A and B are coupled by two connections with streams S_A and S_B. By cutting off, say, the second stream connection, see Fig. 13, and setting some initial value for S_B the system can be solved sequentially (first A then B), and one obtains an output stream $\phi(S_B)$ which leaves B for A.

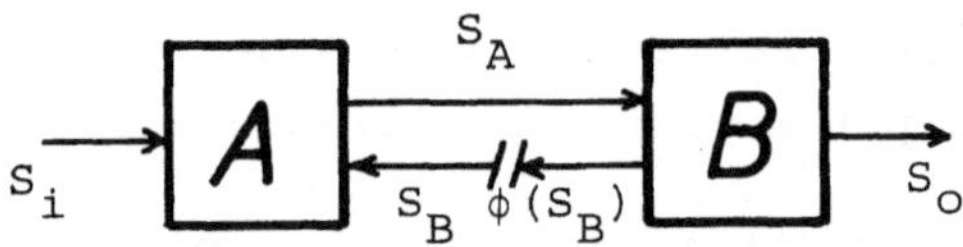

Fig. 13

Therefore, the problem reduces to the fixed point equation

$$S_B = \phi(S_B),$$

which can be solved iteratively, by, e.g., successive approximation techniques, Newton's method or quasi-Newton methods. For general subsystems, a set of tear streams must be introduced to allow a sequential calculation. (At least one edge of each elementary circuit must be cut off, for an efficient method, see [5].) Then the problem reduces to a multidimensional fixed point equation.

We conclude this paper by studying the following "real world" problem:

7   <u>Example</u> (Industrial plant for producing oxygen from air)

The (main part of the) plant consists of 23 devices, (see Fig. 14 for the associated flowsheet), namely 1o splitters (Nos 1,5,9,1o,13,14,15,16,21,23), 2 mixers (Nos 2,19), 2 coolers (Nos 4,17), 1 compressor (No 3), 1 turbine (No 6), 4 heat exchangers (Nos 7,11,18,22) and 3 separation columns (No 8: 22 plates, No 12: 2o plates, No 2o: 39 plates).

It is easy to see that the plant can be partitioned into two subsystems, marked by the broken lines in Fig. 14. The first subsystem is trivial, it consists of only one simple device and is easily solved. No further partitioning of the remaining plant is possible. In order to solve this subsystem, eight tear streams are introduced, denoted by the symbol $\not\vdash$ in Fig. 14. Then the problem reduces to the fixed point equation

$$S_i = \phi_i(S), \quad i = 1,\ldots,8$$

with $S = (S_1,\ldots,S_8)$. After initializing the tear streams, the system can be solved sequentially. For example, by starting with device 1o and 15 and proceeding with device 11 and 9, the liquid feed of column 8 can be calculated. This specifies, together with the tear stream S_8, the input data for column 8. Then the techniques of Sections 4 and 5 are used to obtain the output of the column. In Fig. 15 typical input data are specified and the results of the calculation are shown.

By continuing this process one eventually obtains values for $\phi_i(S)$, $i = 1,\ldots,8$ with $S = (S_1,\ldots,S_8)$, which can be used as new starting values for the tear streams, and so on. (Besides this simple approximation technique for solving the fixed point equation, other methods, in particular Newton-like methods are studied.)

Final results for this plant are not yet available. Work is still going on on this project.

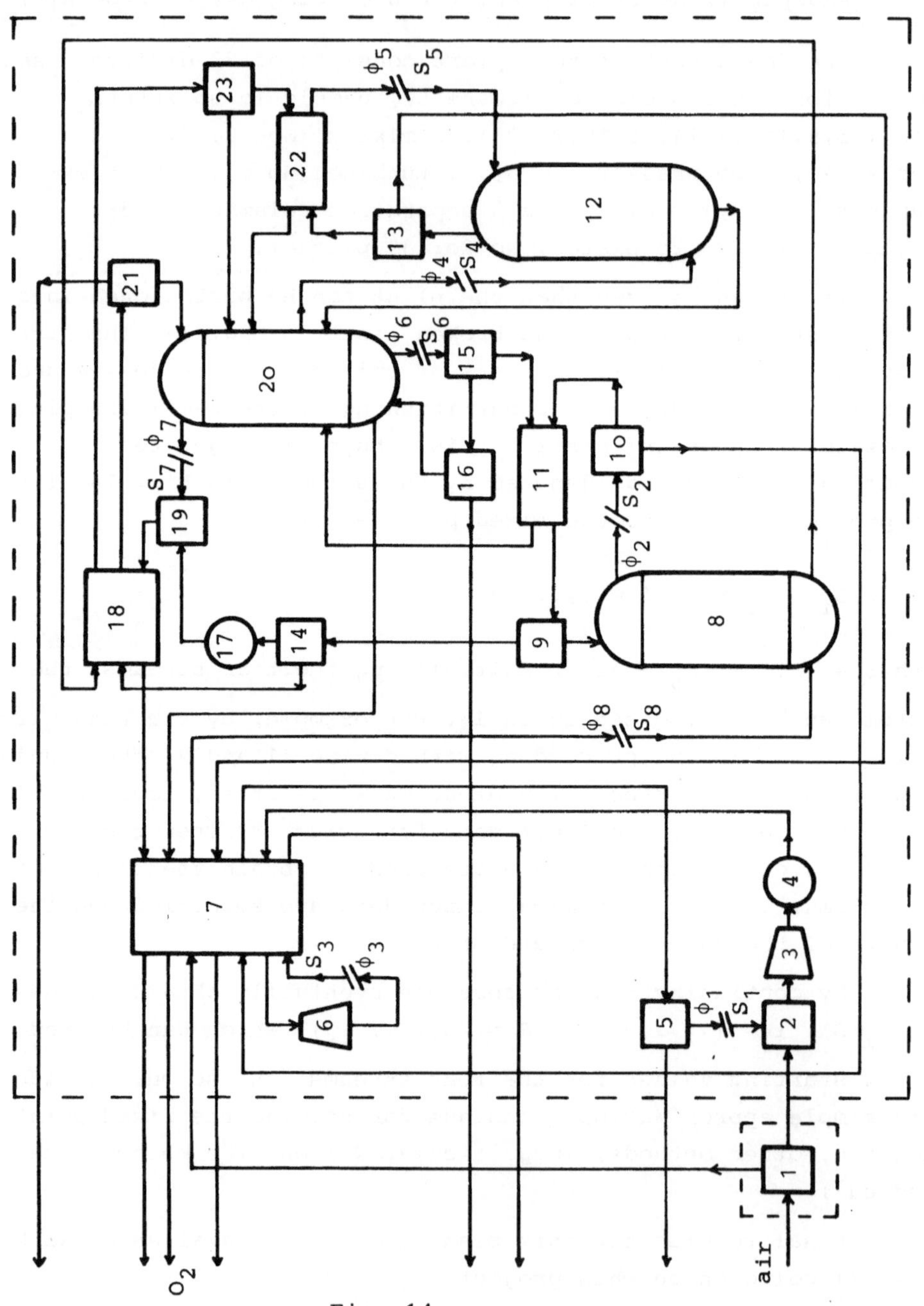

Fig. 14

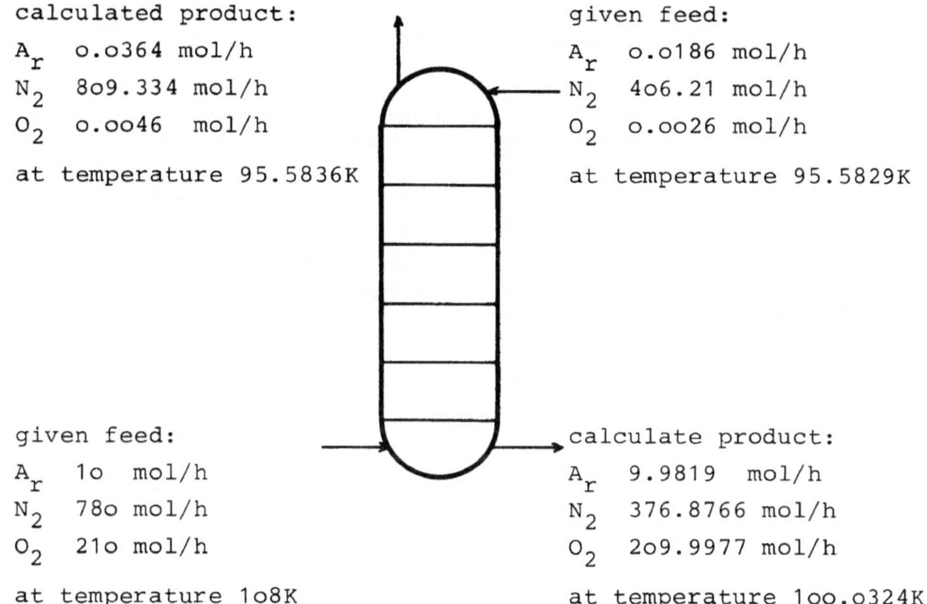

Fig. 15. Separation column No 8
(pressure linearly increases between
5.5 bar (top) and 5.7 bar (bottom))

<u>Acknowlegement:</u>

We would like to thank F.Kokert (VOEST-ALPINE AG) for his valuable cooperation.

<u>References:</u>

[1] Fredenslund, A.; Gmehling, J.; Rasmussen, P.: Vapor-Liquid
    Equilibrium Using UNIFAC. Amsterdam: Elsevier 1977

[2] Hackl, J.; Wacker, Hj.; Zulehner, W.: An efficient step size
    control for continuation methods. BIT 2o (198o) 475-485

[3] King, C.J.: Separation Processes. 2nd ed. New York: McGraw-
    Hill 198o

[4] Kubiček, M.; Hlavaček, V.; Procháska, F.: Global modular
    Newton-Raphson technique for simulation of an interconnected
    plant applied to complex rectification columns. Chem.Engng.
    Sci 31 (1976) 277-284

[5] Pho, T.K.; Lapidus, L.: An optimum tearing algorithm for
    recycle systems. AiChE J. 19 (1973) 117o-1181.

[6]   Schubert, L.K.: Modification of a quasi-Newton method for
      nonlinear equations with a sparse Jacobian. Math. Comp. 24
      (197o) 27-3o

[7]   Tiernan, J.C.: An efficient search algorithm to find the
      elementary circuits of a graph. Comm.ACM 13 (197o) 722-726

[8]   Wacker, Hj. (ed.): Continuation Methods. New York.
      Academic Press 1978

[9]   Westman, K.R.; Lucia, A.; Miller, D.C.: Flash and
      distillation calculations by a Newton-like method. Comp.
      Chem.Engng. 8 (1984) 219-228

OPTIMIZATION OF SYSTEMS OF HYDRO ENERGY POWER PLANTS

Wolfgang Bauer, Ewald H.Lindner and Hansjörg Wacker

1   Introduction

1.1   Hydro Energy Production in Austria

In most European countries energy production is mainly
done by caloric and/or nuclear plants whereas hydro energy
plants are of less importance. Hydro energy production in the
FRG, for instance, runs up to about 5 %. The following figures
demonstrate that in Austria things are quite the opposite.

year	hydro energy [GWh]	caloric energy [GWh]	imports [GWh]	total [GWh]
1960	1o.345	2.593	851	13.789
1970	19.296	6.22o	1.6o5	27.121
198o	27.o15	9.342	3.492	39.849
1982	28.63o	8.926	3.4o8	4o.964
1983	28.295	8.9o4	4.681	41.88o

Table 1: Production of energy in Austria

Although the ratio hydro energy production versus total energy
production decreased from 75 % (196o) to 68 % (1983) river power
plants and storage power plants still are the main suppliers of
electric energy in Austria. In 1983 river power plants produced
19432 GWh whereas storage power plants produced 8863 GWh. Nevertheless, storage power plants are more important as they allow to
transfer energy production with respect to time quite easily. The
natural period of a storage plant - i.e. the time-interval the
full reservoir is emptied and refilled again - may range from
some hours to one year. Most storage plants also allow for pumping

which is done when cheap energy is available. Energy from
storage plants is mainly used for two purposes:
i) for supplying energy in high winter and early spring
ii) for satisfying peak power demands.
Peak power, in addition, can be exchanged at a profitable rate
against base load with neighbouring countries.

1.2   Optimization of the Gosau Power Plant Chain

In 1982 the local electric power plants company in Upper
Austria, OKA, asked for advice concerning the optimal operating
of two of their systems: Gosau - Gosauschmied - Steeg and
Partenstein. We made a first report which showed for both cases
that the value of the production might be increased by a suitable
control of the systems. Our study was backed up by numerous
results we got from our students in problem seminars. The
company got convinced and decided to start with Gosau. Partenstein
was treated also in problem seminars but at a lower degree of
intensity. Nevertheless, a substantial increase of the production
was possible (5 % - 1o %). The Gosau system was solved in the
years 1982 until 1985. The main difficulties consisted
i) in getting correct input data, e.g. influx, seepage losses,
    efficiencies, and
ii) in coping with the high dimensionality of the resulting
    optimization problem.

In this paper we describe how the problem was solved by
quite different methods: classical dynamic programming,
variational techniques for simplified models, a newly developed
decomposition-convexification technique proposed by Gfrerer and
a nonlinear programming technique to cope with peak power demands.
To solve the problem of the Gosau seepage losses quite a few
investigations were done during the last decades. To reduce the
dimensionality of the problem we succeeded to split the problem
into three independent parts. Part I, lowering the reservoir
during autumn and early winter, was put into practice fully in
1983. Though this year was very dry a substantial increase of
production (resp. income) was gained. Part II and III were
finished in 1985, again increasing the result.

2   Basic Information on Hydro Energy Production – a First Solution

In this Chapter we will give some basic information on hydroelectric storage plants and present a solution technique for a simplified model.

2.1   Hydroelectric Storage Plants

Figure 1 gives the scheme of a hydroelectric storage plant.

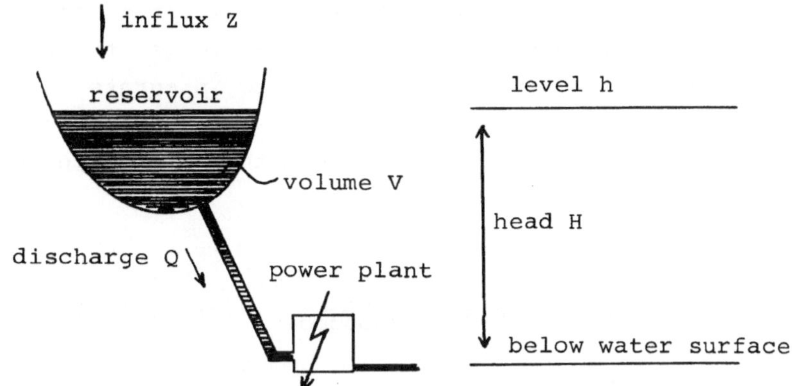

Figure 1: Scheme of a hydroelectric storage plant

The influx Z is stored in the reservoir. The theoretical power of the power plant is

$$P_{Th}[W] = g[m/s^2]\rho[kg/m^3]H[m]Q[m^3/s]$$

where $g = 9.81$ and $\rho = 1000$ is the acceleration due to gravity, and the density of water, resp. The power, which is available in reality, is given by a reduction of the theoretical power. This reduction is described by the efficiency. In our case, the efficiency is a nonlinear function depending on H and Q. We denote it by $\eta(H,Q)$. Hence the real power P is

$$P = P_{Th}\cdot\eta(H,Q) \qquad [W]$$

(Details about η are given in section 6.3).
The volume V of the reservoir may be described as a function of the head H

$$V = C(H).$$

Usually, C is a convex strictly increasing smooth function. Hence, there exists the inverse function, denoted by f. We get $H = f(V)$, which gives the head in dependence on the volume. The variation in time of the volume is described by the continuity equation

$$\frac{dV}{dt} = Z - Q$$

where the values of Z and Q are the sum of different influx resp. discharge components (see section 4). To get the value E of the energy produced in [0,T], we have to integrate the power. As we may not expect, that P resp. Z or Q are constant during the whole interval [0,T], we have to respect the dependence on t. We get:

$$E = 1/3600000 \int_0^T P(t)dt = \frac{g \cdot \rho}{3600000} \int_0^T H(t)Q(t)\eta(H(t),Q(t))dt=$$

$$= const. \int_0^T f(V(t))Q(t)\eta(f(V(t)),Q(t))dt \quad [kWh]$$

Note: 1 kWh = 3600000 Ws

Our aim is to maximize the income G from selling the energy produced in [0,T]. We get G by integrating the product of P(t) and the tariff a(t).

$$G = const. \int_0^T a(t)P(t)dt \quad [Austrian\ Shillings]$$

(We denote the value of G by "rated energy" production). The tariff is a step function, which varies seasonally, weekly and daily. For demonstration we present the tariff a(t) – in Austrian Shillings per kWh – which is valid now (April 1987) – Table 2.

Table 2: Tariff-values [Austrian Shillings/kWh]

	Summer May until August	Transition Period April, September	Winter October until March
high tariff	o.433	o.546	o.571
low tariff	o.382	o.489	o.489

Note: 12,68 Austrian Shillings = 1 US$ (March 31st, 1987)

Constraints: For most applications there are both box constraints on $Q(t)$ and $V(t)$ resp. $H(t)$ and some additional mostly
linear restrictions.

2.2 The Model (M)

Using $Q(t)$ as the control variable and $V(t)$ as the state
variable we get the following model for our optimization problem:

Find a control $Q(t)$ and a state $V(t)$, $t \in [0,T]$ such that

$$G = \text{const.} \int_{0}^{T} a(t)f(V(t))Q(t)\eta(f(V(t)),Q(t))dt$$

is a maximum, subject to the continuity equation

$$\frac{dV(t)}{dt} = Z(t) - Q(t), \quad V(o) = V_o, \quad V(T) = V_T$$

and the box constraints on V and Q:

$$V_{min} \leq V(t) \leq V_{max} \wedge Q_{min} \leq Q(t) \leq Q_{max}$$

2.3 Application of Necessary Conditions

In this section we discuss the possibility to transform
(M) into an equivalent finite-dimensional optimization problem
under the following additional assumptions:

$$Q_{min} < Z(t) < Q_{max} \quad \text{for all } t \in [0,T]$$

i.e. during the times of no production the reservoir level rises and vice versa falls during periods of maximum production.

$f'(V) > 0$ (see 2.1)

The efficiency η is constant.

$[0,T] = 1$ week.

$Q_{min} = 0$.

These assumptions are related to several real world storage power plants (e.g. power plant Gosauschmied or power plant Partenstein), see Gfrerer [5].

2.3.1 Necessary Conditions for an Optimum.

Let $(V_*(t),Q_*(t)) \in C[0,T] \times L_\infty[0,T]$ be a solution of (M). Then there holds: There exist real numbers 1, $\lambda_o \geq 0$; a function ψ and nonnegative measures μ_1 resp. μ_2 with support

$\bar{V}: = \{t \in [0,T] | V_*(t) = V_{max}\}$ resp. $\underline{V}: = \{t \in [0,T] | V_*(t)=V_{min}\}$, not all zero, so that

$$-\psi(t) = -1-\lambda_o \int_t^T a(\tau)f'(V_*(\tau))Q_*(\tau)d\tau + \int_t^T 1d\mu_1(\tau)$$

$$- \int_t^T 1d\mu_2(\tau)$$

and

$$(\psi(t)-\lambda_o a(t)f(V_*(t)))(Q-Q_*(t)) \geq 0$$

for all Q satisfying $0 \leq Q \leq Q_{max}$ and almost all $0 \leq t \leq T$. If we denote

$$s(t): = \psi(t)-\lambda_o a(t)f(V_*(t))$$

$$\bar{Q}: = \{t \in [0,T] | Q_*(t) = Q_{max}\} \text{ and}$$

$$\underline{Q}: = \{t \in [0,T] | Q_*(t) = Q_{min}\}$$

we can reformulate the conditions as follows

$$t \in \bar{Q} \implies s(t) \leq 0$$

$$t \in \underline{Q} \implies s(t) \geq 0$$

$$t \notin (\bar{Q} \cup \underline{Q}) \implies s(t) = 0$$

It is obvious that the optimal solution is either bang-bang or singular.

2.3.2 <u>An Equivalent Finite-Dimensional Model.</u> For a detailed analysis of the necessary conditions we refer to Gfrerer [5]. We only present the results. Let us denote the points of discontinuity of the tariff $a(t)$ by

$$\tau_k = k.86400 \quad [s] \quad k = 0,\ldots,7$$

$$\xi_k = 57600 + (k-1).86400 \quad [s] \quad k = 1,\ldots,7$$

(i.e. τ_{k-1} resp. ξ_k corresponds to 6 a.m. resp. 1o p.m. for day k, $k = 1,\ldots,7$, $\tau_7 = T$).

Then

$$a(t) = \begin{cases} a_H & t \in \,]\tau_{k-1},\xi_k] \\[2mm] a_L & t \in \,]\xi_k,\tau_k] \end{cases} \qquad k = 1,\ldots,7$$

<u>Theorem:</u>

(i) $t \in \bar{V} \cup \underline{V} \cup \bar{Q} \cup \underline{Q}$ a.e. for $t \in [0,T]$

(ii) $\underline{V} \subset \{\xi_i \mid i=1,\ldots,7\}$

(iii) $s(t_o) < 0 \implies s(t) < 0$ in $]t_o,\bar{t}]$

       $s(t_o) > 0 \implies s(t) > 0$ in $]\underline{t},t_o]$

       $s(t_o) = 0 \wedge t_o \notin \bar{V} \implies s(t) < 0$ in $]t_o,\bar{t}]$, $s(t)>0$ in $]\underline{t},t_o[$

       where

       $\bar{t}: = \min\{\xi_i \mid \xi_i \geq t_o\}$ resp. $\bar{t} = T$ for $t_o > \xi_7$

       $\underline{t}: = \max\{\xi_i \mid \xi_i < t_o\}$ resp. $\bar{t} = 0$ for $t_o \leq \xi_1$

(iv) $s(t_o) < 0$ for $t_o \in \,]\xi_i,\tau_i] \implies s(t) < 0$ in $[t_o,T]$

$$s(t_o) > 0 \text{ for } t_o \in \ ]\tau_{i-1}, \xi_i] \Rightarrow s(t) > 0 \text{ in } [0, t_o]$$

This theorem gives already the structure of the optimal solution $(V_*(t), Q_*(t))$ (see Figure 2).

Figure 2: Structure of the optimal solution

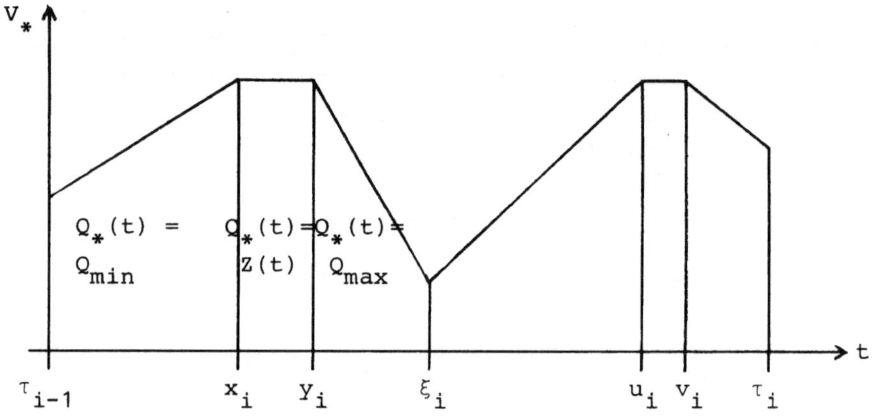

The corresponding values of $V_*(t)$ result from the continuity equation. We get the following optimization problem: Find 28 variables (x,y,u,v) such that

$$a_H \sum_{i=1}^{7} \left[\int_{\tau_{i-1}}^{x_i} Q_{min} \ f(V(\tau_{i-1}) + \int_{\tau_{i-1}}^{t} (Z(\tau)-Q_{min})d\tau) dt \right.$$

$$+ \int_{x_i}^{y_i} Z(t) f(V(x_i)) dt + \int_{y_i}^{\xi_i} Q_{max} \ f(V(x_i) +$$

$$\left. + \int_{y_i}^{t} (Z(\tau)-Q_{max})d\tau) dt \right] +$$

$$+a_L \sum_{i=1}^{7} \left[\int_{\xi_i}^{u_i} Q_{min}\, f(V(\xi_i) + \int_{\xi_i}^{t} (Z(\tau)-Q_{min})d\tau)dt + \right.$$

$$+ \int_{u_i}^{v_i} Z(t)f(V(u_i))dt + \int_{v_i}^{\tau_i} Q_{max}\, f(V(u_i) +$$

$$\left. + \int_{v_i}^{t} (Z(\tau)-Q_{max})d\tau)dt \right]$$

is a maximum with

$$V(x_i): = V_o + \sum_{j=1}^{i} \int_{\tau_{j-1}}^{x_j} (Z(t)-Q_{min})dt +$$

$$+ \sum_{j=1}^{i-1} \left[\int_{y_j}^{\xi_j} (Z(t)-Q_{max})dt + \int_{\xi_j}^{u_j} (Z(t)-Q_{min})dt + \right.$$

$$\left. + \int_{v_j}^{\tau_j} (Z(t)-Q_{max})dt \right]$$

$$V(u_i): = V_o + \sum_{j=1}^{i} \left[\int_{\tau_{j-1}}^{x_j} (Z(t)-Q_{min})dt + \int_{y_j}^{\xi_j} (Z(t)-Q_{max})dt + \right.$$

$$\left. + \int_{\xi_j}^{u_j} (Z(t)-Q_{min})dt \right] + \sum_{j=1}^{i-1} \int_{u_j}^{\tau_j} (Z(t)-Q_{max})dt.$$

The constraints are given by

$$\tau_{j-1} \le x_j \le y_j \le \xi_j \le u_j \le v_j \le \tau_j, \quad j = 1,\ldots,7$$

$$V(x_j) \le V_{max} \qquad\qquad j = 1,\ldots,7$$

$$V(u_j) \le V_{max} \qquad\qquad j = 1,\ldots,7$$

$$V(\xi_j) \ge V_{min} \qquad\qquad j = 1,\ldots,7$$

$$V(\tau_7) = V_T$$

where

$$V(\tau_j): = V(u_j) + \int_{v_j}^{\tau_j} (Z(t)-Q_{max})dt$$

$$V(\xi_j): = V(x_j) + \int_{y_j}^{\xi_j} (Z(t)-Q_{max})dt.$$

Especially, there holds $x_j = \tau_{j-1}$ and $v_j = \tau_j$ if $V_o = V_T = V_{max}$.

2.4 Numerical Solution

The problem is of the following type

$$P: \min\{f_o(x) \,|\, f_j(x) \leq 0, \; i = 1,\ldots,m, \quad x \in \mathbb{R}^n\}$$

To get a starting value for a local iteration procedure, we use the continuation method.

$$P(t): \min\{h_o(x,t) \,|\, h_j(x,t) \leq 0, \; j = 1,\ldots,m, \; x \in \mathbb{R}^n\}$$

$$t \in [0,1]$$

Example:

$$h_o(x,t): = tf_o(x)+(1-t)\|x-x_o\|_2^2$$

$$h_j(x,t): = f_j(x)+(t-1)|f_j(x_o)|$$

We assume that there exists a locally unique solution of $P(t)$: $x(t) \in C[0,1]$. To reduce the numerical effort we use an active index set strategy. With $I(t): = \{j\in\{1,\ldots,m\} \,|\, h_j(x,t)=0\}$ only the following equality constrained problems have to be solved

$$P^I(t): \min\{h_o(x,t) \,|\, h_j(x,t) = 0, \; j \in I(t)\}$$

Under suitable assumptions - $h_i(x,t)$ sufficiently smooth, $\{\nabla h_j(x,t)\}_{j=1}^m$ linearly independent, a strong second order condition for $P^I(t)$, a finite number of critical points, i.e. points where $I(t)$ changes - we have to solve the following nonlinear system

$$\nabla_x h_o(x(t),t) + \sum_{j \in I_o} u_j(t) \nabla_x h_j(x(t),t) = 0$$

$$h_j(x(t),t) = 0 \qquad j \in I_o$$

(u: Lagrange multiplier)

(I_o: interval where I(t) remains constant)

The nonlinear system is solved by classical homotopy techniques, for instance using a p-c method with Euler predictor and Newton corrector. In addition one has to determine the new active index set after passing a critical point. This algorithm is numerically feasible in the sense of Avila and converges quadratically. For details see Gfrerer et al [8], [9].

We present a numerical result (Figure 3) considering the case $V_o = V_T = V_{max}$ and an intermediate tariff (i.e. during the weekend the tariff changes only on Saturday 1 p.m.). The data we use belong to the storage power plant Partenstein (see [1o] this volume).

Figure 3: Numerical example

i[day]	1	2	3	4	5	6
y_i[h]	9.51	33.62	57.4	81.1	98.3	12o
u_i[h]	24	48	72	96	12o	141.7

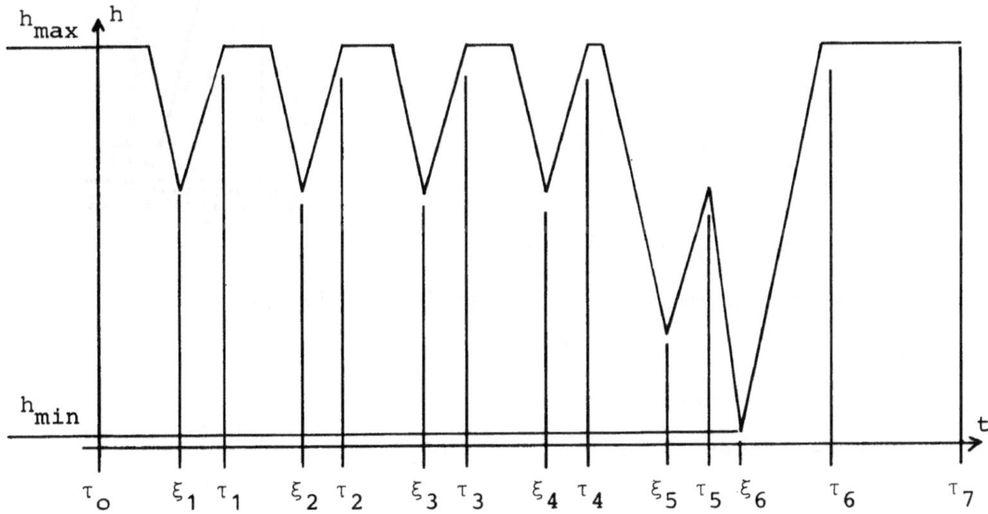

Remark: The structure of the solution (Figure 3) is used for the numerical model of the optimization of the Gosau System during Phase II.

3   Description of the System Gosau

3.1   The System

The system consists of three serially connected storage power plants, each with a reservoir of different capacity.

Figure 4 presents a schematical description of the system (for details see Barwig/Peßl [1]).

Figure 4: Schematical description of the system Gosau

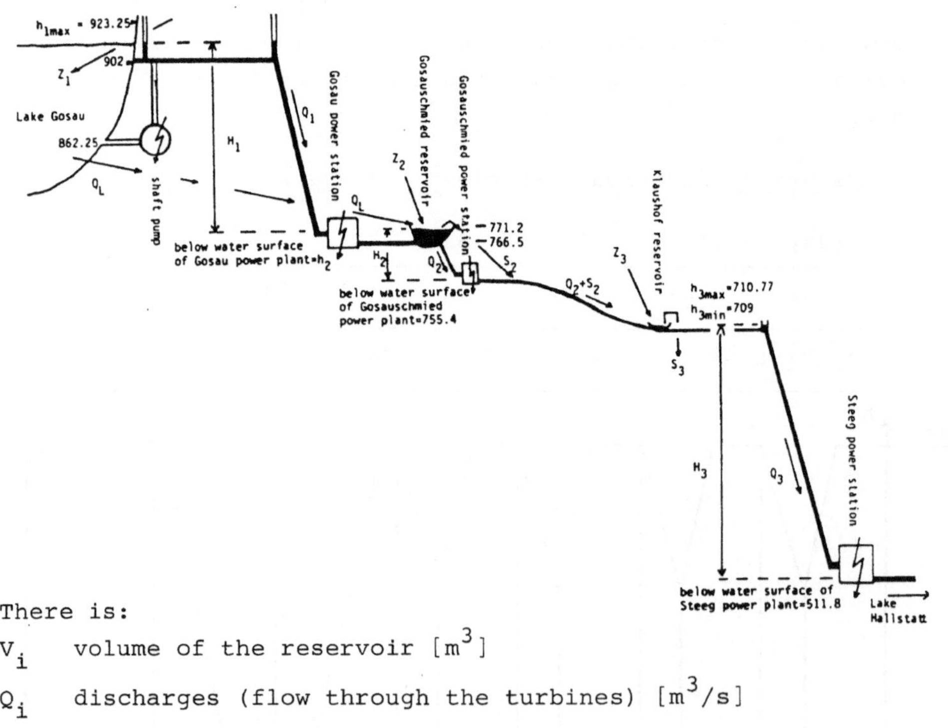

There is:

V_i volume of the reservoir $[m^3]$

Q_i discharges (flow through the turbines) $[m^3/s]$

H_i head $[m]$

h_i height above Adria $[m]$

Q_L seepage losses of Lake Gosau $[m^3/s]$

S_i spillage water $[m^3/s]$

Z_i influx $[m^3/s]$

subscript i	power plant	reservoir	capacity
1	Gosau	Lake Gosau	1 year
2	Gosauschmied	Gosauschmied	1 day
3	Steeg	Klaushof	1-2 hours

3.2   The Problem

Our aim is to maximize the annual "rated energy" production (see Chapter 2) of the whole system, respecting all restrictions, of course.

Lake Gosau is situated in a karst region of the Austrian Alps, hence there are heavy seepage losses Q_L (up to 1.5 m^3/s) depending on the level or volume, resp. The influx and the seepage losses of Lake Gosau are subterraneous and cannot be measured directly (for a determination of Z_1 and Q_L see Sect.6). Q_L is lost for production at Gosau but adds to the influx of Gosauschmied. A special shaft pumping system has to be activated for exploiting Lake Gosau below a level of 9o2.5 m above Adria (= m a.A.). Spillage water is possible for Gosauschmied and Klaushof. For tourism the level of Lake Gosau during summer must be at least 915 m a.A., and 92o m a.A. during July and August. In Steeg a power of 1.5 x $1o^6$ [W] has to be produced every day from 6 a.m. until 1o p.m. because of a supply contract with the ÖBB (Austrian Railway Company).

3.3   Management of the System

Production is performed mostly at Steeg. Nevertheless the system is controlled mainly by Gosau because of the small capacity both of Gosauschmied and Klaushof.

The operation of the system can be split into three independent parts (see Figure 5).

Figure 5. Scheme of the level of Lake Gosau

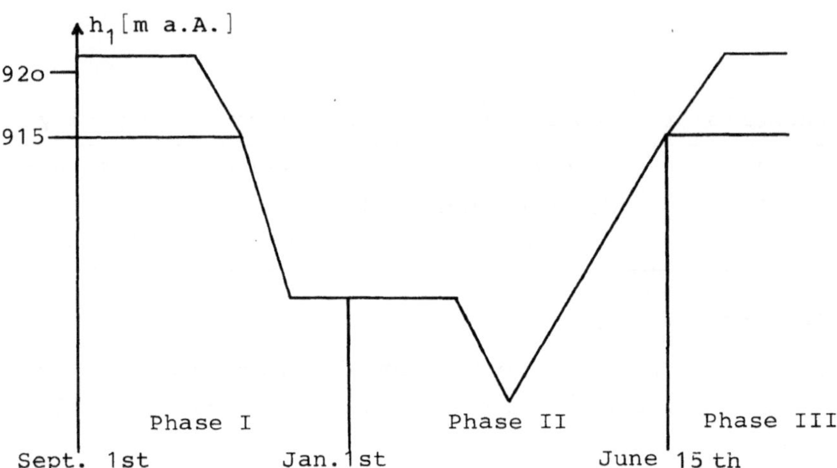

Phase I: The earliest starting point for sinking Lake Gosau in autumn is fixed at September 1st because of tourism. The results of the optimization show that this can be done without loss of generality. The sinking process ends also w.l.o.g at December 31st at the level of 9o2.5 m a.A. Below that level the seepage losses remain constant. Additionally, the shaft pump system must be activated below that level.

Phase II: First the level of 9o2.5 m is kept fixed because of purposes of reserve - only the natural influx is used for production. Then the sinking process is continued using the shaft pump system where the starting point, the deepest level and the intensity of the sinking process are determined by the optimization. Phase II ends at June 15th when a level of 915 m a.A. must be reached both by the natural influx coming from the Dachstein glacier and by pumping activities (the water is pumped back from the reservoir Gosauschmied to Lake Gosau).

Phase III: Lake Gosau is stored up to a level of 92o m a.A. and kept at this height.

Therefore, our problem may be split into three independent sub-problems defined by the phases I, II, III. While Phase III is trivial the two other problems lead to nonlinear problems of optimal control.

4   The Mathematical Model (M)

4.1   Basic Considerations about (M)

As in part 2 we choose
$Q(t) = (Q_1(t),Q_2(t),Q_3(t))^T \in L_\infty^{(3)}[O,T]$ for the control variable
and $V(t) = (V_1(t),V_2(t),V_3(t))^T \in C^{(3)}[O,T]$ for the state
variable. The continuity equations describing the water balance
in the reservoirs are written in integral form now, because
$V(t)$ need not be continuously differentiable.

The objective G (see Section 2) will be now

$$G = \text{const.} \int_o^T a(t) \sum_{i=1}^3 P_i(t)\,dt$$

where $P_i(t)$ is the electric power produced of plant i at time t.
Let $F(V)$ be given by

$$F(V): = \text{diag}(f_1(V_1)-f_2(V_2),\ f_2(V_2)-755.4,\ f_3(V_3)-511.8) =$$

$$=: (f_{ij}(V))$$

and η resp. φ be vector functions defined by

$$\eta(V,Q) := (\eta_1(f_{11}(V),Q_1),\ \eta_2(f_{22}(V),Q_2),\ \eta_3(f_{33}(V),Q_3))^T$$

$$\varphi(V,Q,t) = \begin{pmatrix} \varphi_1(V,Q,t) \\ \varphi_2(V,Q,t) \\ \varphi_3(V,Q,t) \end{pmatrix} :=$$

$$:= \begin{pmatrix} Z_1(t)-Q_1(t)-Q_L(V_1(t)) \\ Z_2(t)-Q_2(t)+Q_1(t)+Q_L(V_1(t))-S_2(V_2(t)) \\ Z_3(t)-Q_3(t)+Q_2(t-\tau_o)+S_2(V_2(t-\tau_o))-S_3(V_3(t)) \end{pmatrix}$$

Remark:

- The diagonal elements $f_{ii}(V)$ give the heads H_i

- η_i are the efficiencies of storage power plant i

- the components φ_i describe the variation of V_i at time t (cf. 2.1, continuity equation)

- the water coming from the reservoir Gosauschmied needs approximately one hour (= τ_o) to arrive at the reservoir Klaushof.

4.2   The Model (M)

Find a vector function Q such that for V_o and V_T ($\in \mathrm{I\!R}^3$) given

$$(M1) \quad const \int_o^T a(t) \sum_{i=1}^3 P_i(t)dt =$$

$$= const. \int_o^T a(t)Q(t)^T F(V(t))\eta(V(t),Q(t))dt$$

is a maximum, subject to

$$(M2) \quad V(t) = V_o + \int_o^t \varphi(V,Q,\tau)d\tau \quad t \in [0,T]$$

$$(M3) \quad V(T) = V_T$$

$$(M4) \quad V_{imin} \leq V_i(t) \leq V_{imax} \quad i = 1,2,3 \quad t \in [0,T]$$

$$V_1(t) \geq \begin{cases} V_{1,915} & I_1 \setminus I_2 \\ V_{1,92o} & \text{for } t \in I_2 \\ V_{1,9o2.5} & I_3 \end{cases}$$

where $I_1 := [\text{June 15th, Oct.15th}]$, $I_2 := [\text{July 15th, August 31st}]$ and $I_3 := [\text{Oct.16th, Dec.31st}]$.

$$(M5) \quad Q_{imin} \leq Q_i(t) \leq Q_{imax} \quad i = 1,2,3 \quad t \in [0,T]$$

(M6) $g \cdot \rho \cdot Q_3(t) f_{33}(V(t)) \eta_3(f_{33}(V(t)), Q_3(t)) \geq 1,5 \times 10^6$

$$\text{for } t \in [6 \text{ a.m.}, 10 \text{ p.m.}]$$

Remark:

(M1) gives the value of the "rated energy" production in the interval $[0,T]$. In any case this objective is nonconcave because of the tariff $a(t)$.

(M2) presents the continuity equations with the initial condition $V(0) = V_o$.

(M3) gives the terminal conditions on V.

(M4) For a discussion of the restrictions on V_1 see Section 3.3.

(M5) The flow of water through the turbines is restricted. We have $Q_{2min} = Q_{3min} = 0$ and $Q_{1min} < 0$ (pumping).

(M6) takes into account the supply contract with the ÖBB (see Section 3.2).

5   Numerical Models

(M) is solved by use of numerical methods for nonlinear optimization. Discretization of the interval $[0,T]$ and assumptions about the control Q reduces the control problem (M) to an finite-dimensional optimization problem.

5.1   Dynamic Programming

We get the optimization problem by
- Discretization of $[0,T]$ into N subintervals

$$[0,T] = [t_o, t_1] \cup]t_1, t_2] \cup \ldots \cup]t_{N-1}, t_N]$$

where the discontinuities of $a(t)$ (= ξ_i, τ_i see 2.3.2) are a subset of $\{t_k\}_{k=0}^N$.
- Assuming that the control Q is constant in each subinterval $]t_{k-1}, t_k]$.

We use Bellmann's principle to solve our optimization problem. To reduce the high dimensionality of the problem, we neglect the storage capacity of the Klaushof reservoir. We may do this, because the capacity of this reservoir is too small to influence the character of the optimal solution.

The Model (DPM)

- The time steps t_k define the levels k of (DPM) $k = 1,\ldots,N$

- $V_i^{(k-1)}$ $i = 1,2$, $k = 1,\ldots,N$ are the state variables at the beginning of step k (As $V_3(t) = \text{const.}$ only $V_1(t)$ and $V_2(t)$ are state variables)

- $Q_i^{(k)}$ $i = 1,2$, $k = 1,\ldots,N$ are the control variables - step k. $Q_3^{(k)}(t)$ is evaluated from

$$Q_2^{(k)} + S_2(V_2(t)) + Z_3(t) - Q_3^{(k)}(t) - S_3(t) = 0.$$

Using classical Dynamic Programming we have to neglect the time delay τ_o.

- The state and control variables are restricted

$$V_{imin} \leqq V_i^{(k)} \leqq V_{imax} \qquad i = 1,2 \qquad k = 1,\ldots,N$$

$$Q_{imin} \leqq Q_i^{(k)} \leqq Q_{imax} \qquad i = 1,2 \qquad k = 1,\ldots,N$$

In addition we have to take into account the restrictions on V_1 given in (M4), and a nonlinear restriction on Q_2 resp. V_2 because of the supply contract with the ÖBB - (M6).

- The state equations are given by the continuity equations

$$V_i^{(k)} = V_i^{(k-1)} + \int_{t_{k-1}}^{t_k} \varphi_i(V(t),Q^{(k)}(t))dt \quad i=1,2, \ k = 1,\ldots,N$$

with $V_i^{(0)} = V_{i_o}$ $i = 1,2$ and $Q^{(k)}(t) = (Q_1^{(k)},Q_2^{(k)},Q_3^{(k)}(t))^T$

- Terminal condition $V_i^{(N)} = V_{i_T}$ $i = 1,2$

- Objective:

$$\text{const.} \ \sum_{k=1}^{N} a(t_k) \int_{t_{k-1}}^{t_k} Q^{(k)}(t)F(V(t))\eta(V(t),Q^{(k)}(t))dt = \text{Max!}$$

Numerical Solution

(DPM) is used to solve (M) during Phase I. A reduction of the interval of possible state variables using upper and lower bound techniques (see [3]) led to an acceptable amount of CPU-time.

Figure 6 presents the optimization results of the plant Gosau for Phase I in 1972 and in 1974.

Figure 6.

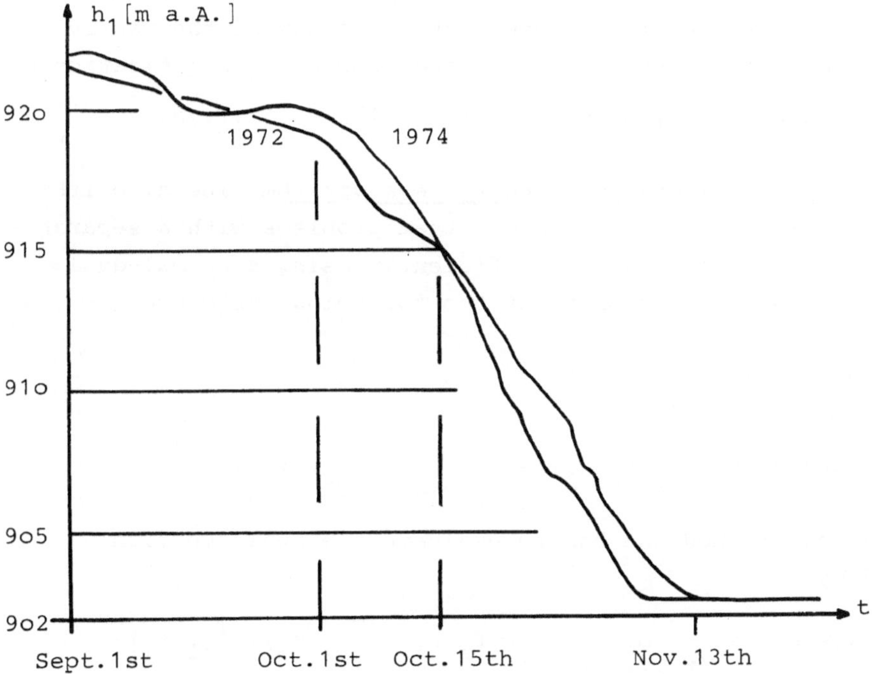

An application to Phase II is possible but the numerical work increases substantially because of the high number of variables and restrictions. For a solution during Phase II see Section 5.5.

5.2   Decomposition - Convexification Method

In this section we describe the application of a primal-dual method, developed by Gfrerer [6], to a short term model ($[O,T] \leq 1$ week). We study this model to check the sensitivity of the results of the long term model (in 5.1) with respect to both its coarse time grid and some simplifications met there. That is, we now take into account the capacity of the reservoir Klaushof as well as the delay τ_o, and choose a discretization of $[O,T]$ so that $t_k - t_{k-1} \leq 3600$ [s] for all $k \in \{1,\ldots,N\}$.

5.2.1 Basic considerations about the algorithm. The algorithm we will apply, bases on the fact, that problems with a separable structure may be solved very efficiently using a primal-dual method. We consider problems of the following structure

$$\text{minimize} \quad \sum_{i=1}^{q} f_i(x_i)$$

$$\text{subject to} \quad \sum_{i=1}^{q} h_i(x_i) = O, \quad \sum_{i=1}^{q} g_i(x_i) \leq O.$$

Using a dual method one has to evaluate the dual function , defined by

$$\phi(\lambda,\mu) := \min_{(x_1,\ldots,x_q)} \{ \sum_{i=1}^{q} f_i(x_i) + \lambda^T h_i(x_i) + \mu^T g_i(x_i) \}$$

Now the minimization problem decomposes into q separate problems

$$\varphi_i(\lambda,\mu) := \min_{x_i} \{ f_i(x_i) + \lambda^T h_i(x_i) + \mu^T g_i(x_i) \}$$

which are of smaller dimension than the original problem. Hence, they can be solved rather efficiently. But the application of primal-dual methods is convenient for convex optimization problems only. Now, our optimization problem is neither separable nor convex. However, a suitable modification of the numerical model of (M) will lead to an equivalent model, which belongs to the required class of separable and convex minimization problems.

5.2.2 <u>The Model.</u> For ease of presentation we confine ourselves to describe the model of power plant Gosau, only. We omit the index 1. The variables are the volumes $V^{(k)} := V(t_k)$, now. We assume V to be a linear function in $]t_{k-1}, t_k]$, $k = 1, \ldots, N$ and a continuous piecewise linear function in $[0,T]$ (i.e. we assume $Q(t)$, $Z(t)$ and $Q_L(V(t))$ to be constant for

$$t \in]t_{k-1}, t_k] : Q^{(k)}, \; Z^{(k)} \text{ and } Q_L^{(k)} := Q_L(\frac{V^{(k-1)} + V^{(k)}}{2}), \; k=1,\ldots,N).$$

The efficiency η is assumed to be constant. We get the model:

Find (N-1) variables $V^{(k)}$, $k = 1, \ldots, N-1$ such that the "rated energy" production

$$\sum_{k=1}^{N} E_k(V^{(k-1)}, V^{(k)}) := \text{const.} \sum_{k=1}^{N} a(t_k) (\frac{V^{(k-1)} - V^{(k)}}{t_k - t_{k-1}} +$$

$$+ Z^{(k)} - Q_L^{(k)}) \int_{t_{k-1}}^{t_k} f(V^{(k-1)} + \frac{V^{(k)} - V^{(k-1)}}{t_k - t_{k-1}}(t - t_{k-1})) dt$$

is a maximum,

subject to

$$V^{(0)} = V_o, \; V^{(N)} = V_T$$

$$V_{min} \leq V^{(k)} \leq V_{max} \qquad k = 1, \ldots, N-1$$

$$Q_{min} \leq \frac{V^{(k-1)} - V^{(k)}}{t_k - t_{k-1}} + Z^{(k)} - Q_L^{(k)} \leq Q_{max} \qquad k = 1, \ldots, N$$

Note: $\dfrac{V^{(k-1)} - V^{(k)}}{t_k - t_{k-1}} + Z^{(k)} - Q_L^{(k)} = Q^{(k)}$ by the continuity equation.

This problem is neither convex nor separable. To obtain separability we introduce new variables $\underline{V}^{(k)}$, $\bar{V}^{(k)}$, $k=1,\ldots,N-1$

with:

$\underline{V}^{(k)}$ is the volume at the end of interval k: $]t_{k-1},t_k]$ and

$\bar{V}^{(k)}$ is the volume at the beginning of interval k+1: $]t_k,t_{k+1}]$.

Now we get the following separable model (MS):

(MS1) minimize $-\displaystyle\sum_{k=1}^{N} E_k(\bar{V}^{(k-1)},\underline{V}^{(k)})$

subject to $\bar{V}^{(o)} = \underline{V}_o,\quad \underline{V}^{(N)} = V_T$

$$V_{min} \leq \bar{V}^{(k-1)} \leq V_{max} \qquad k = 2,\ldots,N$$

(MS2) $\qquad V_{min} \leq \underline{V}^{(k)} \leq V_{max} \qquad k = 1,\ldots,N-1$

$$Q_{min} \leq \frac{\bar{V}^{(k-1)}-\underline{V}^{(k)}}{t_k-t_{k-1}} + z^{(k)} - Q_L\left(\frac{\bar{V}^{(k-1)}+\underline{V}^{(k)}}{2}\right) \leq Q_{max}$$

$$k = 1,\ldots,N$$

To get equivalence between (M) and (MS), we have to add the equality constraints

(MS3) $\underline{V}^{(k)} = \bar{V}^{(k)} \qquad k = 1,\ldots,N-1$

(continuity of V). (MS) is separable but convexity is still hurt in the objective and in the Q-constraints, because of the tariff and of Q_L, resp. The convexification of the Q-constraints is obtained by linearization of Q_L with respect to V. This approximation of Q_L is acceptable for our short term optimization.

The objective is convexified by adding a sufficiently convex term depending on positive scalars $\bar{c}^{(k)}$, $\underline{c}^{(k)}$ and on the variables $\bar{Y}^{(k)}$, $\underline{Y}^{(k)}$:

$$E_k^{(*)}(\bar{V}^{(k-1)},\underline{V}^{(k)},\bar{Y}^{(k-1)},\underline{Y}^{(k)}) := $$

$$-E_k(\bar{V}^{(k-1)},\underline{V}^{(k)}) + \bar{c}^{(k-1)}(\bar{Y}^{(k-1)}-\bar{V}^{(k-1)})^2 + \underline{c}^{(k)}(\underline{Y}^{(k)}-\underline{V}^{(k)})^2$$

Now we add the equality constraints to the objective by duality:

$$\phi(Y,\lambda) := \min\{\sum_{k=1}^{N} E_k^{(*)} + \sum_{k=1}^{N-1} \lambda_k (\bar{v}^{(k)} - \underline{v}^{(k)}) \mid (\bar{v}^{(k)}, \underline{v}^{(k)})$$

subject to (MS2)}

To get the solution of the optimal problem we have to solve the unconstrained Minimax problem

$$\min_{Y\in\mathbb{R}^{2N-2}} \quad \max_{\lambda\in\mathbb{R}^{N-1}} \quad \phi(Y,\lambda)$$

To determine $\phi(Y,\lambda)$ N linearly constrained optimization problems

$$(*) \quad \min\{\varphi_k(Y,\lambda) := E_k^{(*)}(\bar{v}^{(k-1)}, \underline{v}^{(k)}, \bar{Y}^{(k-1)}, \underline{Y}^{(k)}) +$$

$$+ \lambda_{k-1}\bar{v}^{(k-1)} - \lambda_k\underline{v}^{(k)} \mid (\bar{v}^{(k-1)}, \underline{v}^{(k)}) \text{ subject to (MS2)}\}$$

of dimension two only have to be solved. This can be done quite efficiently.

The following iteration procedure is globally and super-linearly convergent:

Start: $i := 0$, $Y_i \in \mathbb{R}^{2N-2}$

Step 1:   maximize $\phi(Y_i,\lambda)$ with respect to λ: λ_i

V_i: corresponding solution of (*) when evaluating $\phi(Y_i,\lambda_i)$

Step 2: $Y_{i+1} := V_i$, $i := i+1$,   Goto Step 1

The efficiency of this technique is demonstrated by the following results:

N	Iterations (It)	CPU (IBM 360-155)	Constraints
42	1o3	14"	166
168	172	1'27"	67o
1oo8	739	35'17"	4o3o

5.2.3 <u>Numerical Example.</u> Our problem can be split up both with respect to the Systems Gosau/Gosauschmied and Steeg and to the time intervals as described above. $\phi(Y,\lambda)$ is determined by solving N five-dimensional problems and N four-dimensional problems. For details see Gfrerer [7].
Table 3 gives the results from an one week optimization which belong to the high tariff period.

Table 3: Results - short term optimization

$t[h]$	$Q_1[m^3/s]$	$Q_2[m^3/s]$	$Q_3[m^3/s]$
6	4.3	6.6	6.1
7	4.3	6.6	7.4
.	.	.	.
.	.	.	.
15	4.3	6.6	7.4
16	o.3	5.6	7.4
.	.	.	.
.	.	.	.
2o	–	6.1	7.4
21	–	–	7.4
22	–	–	–

Remarks:

a) One also may dispose of the inequalities (MS2) by duality. Numerical tests, however, show that the method described above is to be preferred.

b) Some care should be devoted how to choose the constants $\bar{c}^{(k)}$, $\underline{c}^{(k)}$. It can be proved that the "amount of convexification" should be as small as possible to accelerate convergence. For details see Gfrerer [4], [6].

c) To solve the minimax problem a Quasi-Newton like technique is applied.

5.3 <u>Daily Optimization of the System Gosau</u>

In 5.2 we do not respect peak power demands. Now we present a model, which takes into account that there are several times during which production is preferred.

5.3.1 <u>The Model.</u> Production is preferred at the following times in the order: from 6 am till 1 pm, from 6 pm till 1o pm, from 1 pm till 6 pm and from 1o pm till 6 am. Therefore, we choose discretization points 6am, 1pm, 6pm, 10pm, when production may start, and 6am next day. As production need not to be sustained over the whole interval of time, stopping points t_{i1}, t_{i2}, t_{i3} were introduced with constraints 6am $\leq t_{i1} \leq$ 1pm $\leq t_{i2} \leq$ 6pm $\leq t_{i3} \leq$ 10pm. During night time (i.e. 10pm till 6am) we used a point t_{i4}, when the amount of discharge may be changed. This time discretization was used for Gosau (i=1), and Gosauschmied (i=2). In Steeg (i=3) we added two further fixed time points (2pm, 7pm), where the amount of discharge may be changed. This time discretization yields the representation $\bar{Q}_i(t)$ of $Q_i(t)$, where we assume $\bar{Q}_i(t)$ to be constant on each interval of time:

$$\bar{Q}_i(t) = \begin{cases} Q_{i1} & \text{for } t \in \quad [6\text{am},t_{i1}] \\ 0 & \qquad\qquad]t_{i1},1\text{pm}]\cup]t_{i2},6\text{pm}]\cup]t_{i3},10\text{pm}] \end{cases} \quad i=1,2,3$$

$$\bar{Q}_i(t) = \begin{cases} Q_{i2} & &]1\text{pm},t_{i2}] \\ Q_{i3} & \text{for } t \in &]6\text{pm},t_{i3}] \\ Q_{i4} & &]10\text{pm},t_{i4}] \\ Q_{i5} & &]t_{i4},6\text{am next day}] \end{cases} \quad i = 1,2$$

and

$$\bar{Q}_3(t) = \begin{cases} Q_{32} & &]1\text{pm},2\text{pm}] \\ Q_{33} & &]2\text{pm},t_{32}] \\ Q_{34} & \text{for } t \in &]6\text{pm},7\text{pm}] \\ Q_{35} & &]7\text{pm},t_{33}] \\ Q_{36} & &]10\text{pm},t_{34}] \\ Q_{37} & &]t_{34},6\text{am next day}] \end{cases}$$

The model is constructed such that the application of the code EØ4VAF of NAGLIB is efficient. EØ4VAF is a sequential augmented Lagrangian method, where subproblems are solved by a quasi-Newton method. Linear constraints are treated as nonlinear ones, but box constraints are treated separately. EØ4VAF transfers inequality into equality constraints by introducing additional slack variables, therefore, the dimension of the internally treated optimization problems may be very high. The dimensionality is reduced using a penalty-like ansatz. The lower bound constraints on V_2 and V_3 are modelled by extending f_{22} and f_{33} for $V_i < V_{imin}$, $i = 2,3$ and by reducing the efficiency function η_2 and η_3 for $V_i < V_{imin}$, $i = 2,3$. For the data under consideration the upper and lower bounds on V_1 were always inactive, so we dropped them a-priori. This results in the following model:

Find 29 variables $(Q_{11},\ldots,Q_{37},\ t_{11},\ldots,t_{34})$ such that the "rated energy" production of the whole system during a day is a maximum
subject to

$$Q_{imin} \leq Q_{ij} \leq Q_{imax} \qquad i = 1,2;\ j = 1,\ldots,5$$

$$Q_B \leq Q_{3j} \leq Q_{3max} \qquad j = 1,\ldots,5$$

$$Q_{3min} \leq Q_{3j} \leq Q_{3max} \qquad j = 6,7$$

(Q_B is the minimum amount of water to produce the required power for the ÖBB).

$$6am \leq t_{11},t_{21},t_{31} \leq 1pm \leq t_{12},t_{22} \leq 6pm$$

$$2pm \leq t_{32} \leq 6pm \leq t_{13},t_{23} \leq 10pm$$

$$7pm \leq t_{33} \leq 10pm \leq t_{14},t_{24},t_{34} \leq 6am\ \text{next day}$$

and $\quad V(6am\ \text{next day}) = V_T$

Up to now the desired order of the production times is not observed in our model. To cope with this problem we choose the starting point in a suitable way (i.e. the starting value is chosen such that the production periods are part of the

preferred production times, respecting the order of these times).
This will do, because the algorithm is locally convergent.

5.3.2 <u>Numerical Example.</u> Figure 7 gives the results of a day
with high resp. low production.

<u>Remark:</u> From numerous numerical examples (made in 5.2 and 5.3)
we get:
- The operation of the power plant system will be sufficiently
  optimal, if one tries to avoid spillage water, at least at
  Klaushof, and to produce energy during the high tariff period,
  mainly.
- For the daily optimization the efficiency functions and the
  seepage losses are of minor importance.
- The reservoir Gosauschmied is used as an auxiliary reservoir
  for Steeg.

Simulation shows that by observing these rules one gets production values which differ by 1 percent to 2 percent from
the optimal value determined in 5.2.
Moreover, we get that the corresponding objective value of the
longterm optimization is just 0.5 percent on the average and
2.5 percent at most below the optimum.
Hence, the assumptions made for the long term optimization are
acceptable.

5.4   <u>Optimization During Phase II</u>

Now we want to find the optimal operation of the System
Gosau during Phase II. The numerical effort, necessary to solve
(M) for Phase II, using one of the models above, would be
enormous. So we searched for an algorithm which solves our
problem with an acceptable amount of CPU-time.

It is known that the method of Dynamic Programming is
very efficient, if the dimension of the state and control
variable is one. Therefore, we searched for equations which
determine the control variables Q_2 and Q_3 without any optimization. We know some properties of the optimal controls Q_2 and
Q_3 (see Chapter 2 and 5.3). Using these properties we can
formulate a number of equalities and inequalities (denoted by

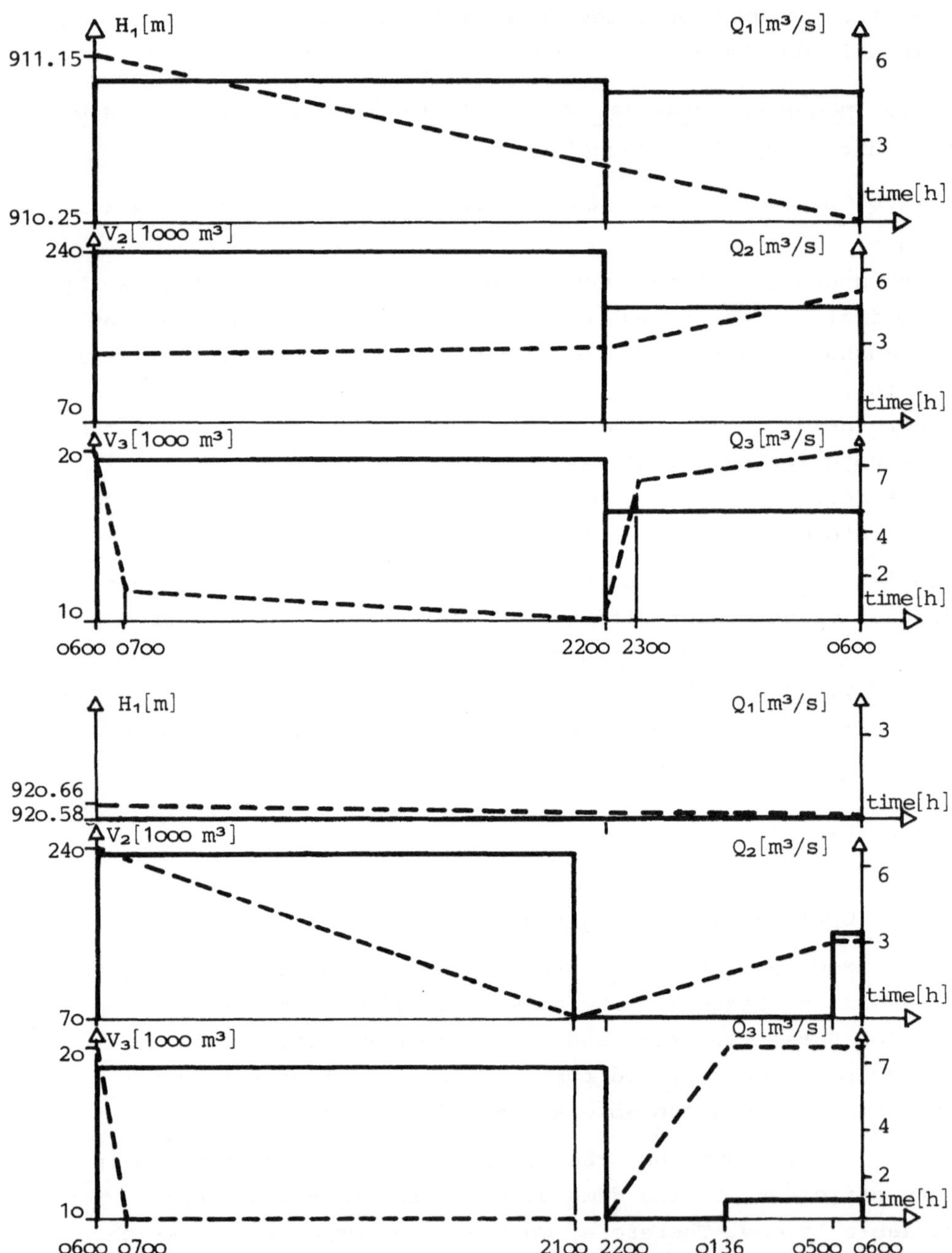

Figure 7: Daily optimization of a day with high, and low production, see top, and bottom, triple diagram, resp. The full, and broken, lines represent the discharges (ordinate on right side), and the content of the reservoirs (ordinate on left side), resp.

(A)), which fix the values of Q_2 and Q_3. We describe (A) verbally.

(A) computes Q_2 and Q_3 such that

- spillage of water is avoided, at least in Klaushof,
- production is mostly done during the high tariff period,
- the volume V_2 reaches a prescribed value at each end of the low tariff period,
- V_3 is constant (we neglect the storage capacity of the Klaushof reservoir),
- the power plants are run near their optimum efficiency;
- all restrictions on Q_2 and V_2 are fulfilled (restrictions on Q_3 are replaced by restrictions on Q_2, see e.g. Section 5.1);
- Q_2 is a step function. The value of Q_2 may change only at those points where the tariff has a jump.

A detailed description of (A) is given in Bauer [3].

Using (A) we may dispose of the control variables Q_2 and Q_3 and of the state variables V_2 and V_3, if Q_1 is known. This fact reduces the numerical effort substantially.

The numerical model is derived from (M) by discretizing the interval [0,T] and assuming that the control is a step function which may have jumps at the discretization points t_k only.

The discretization of [0,T] is done such that for each subinterval $]t_{k-1},t_k]$ holds

$$1 \text{ day} \leq t_k - t_{k-1} \leq 7 \text{ days}$$

<u>Note:</u> The unit of the times t_j is seconds. Other units, like above, are used only if this results in more clearness. We get N = 52 upto N = 365 depending on the choice of the intervals $]t_{k-1},t_k]$. (N is the number of points t_k, see section 5.1).

Now let us define 1o operation modes for the Gosau plant, which will be the control variables of the model below (see Table 4).

Table 4: Definition of the operation modes

OM.number	Definition
1	the level of Lake Gosau is held constant
2	"optimal production" - four hours during the high-tariff period
3	"optimal production" - eight hours during the high-tariff period
4	"optimal production" - twelve hours during the high-tariff period
5	"optimal production" - sixteen hours during the high-tariff period
6	storage of the influx Z_1
7	storage of Z_1 / spillage of water at the Gosauschmied reservoir is avoided by pumping it up to Lake Gosau
8	Gosau pump in operation during the whole low-tariff period
9	OM.8 / spillage of water at the Gosauschmied reservoir during the high-tariff period is pumped up to Lake Gosau
1o	pumping during day and night

Remark: "Optimal production" means: Q_1 is chosen so that the plant works at its optimum efficiency. A constant OM does not imply a constant flow through the Gosau pumping-turbine in $]t_{k-1}, t_k]$. The definition of the OMs reflects the way, how decision is made in practice. The computation of the values of Q_1 for given OM is similar to the determination of Q_2 and is done by an extension of the algorithm (A). (For details see [3]).

5.4.1 The Dynamic Model.

- The times t_k define the end points of step k, k = 1,...,N.
- $V_1^{(k-1)}$, k = 1,...,N is the state variable at the beginning of step k.

- $OM^{(k)} \in \{OM_1,...,OM_{1o}\}$ is the control variable - step k.

From this we can determine the functions Q(t) and V(t) for $t \in]t_{k-1}, t_k]$ by means of algorithm (A). That is, the state equations are evaluated in (A).

- Boundary condition: $V_1^{(N)} = V_{1_T}$.

- Only restrictions on $V_1^{(k)}$ have to be taken into account

$$V_{1min} \leq V_1^{(k)} \leq V_{1max} \qquad k = 1,\ldots,N$$

and

$$V_1^{(N)} \geq V_{1,915}$$

All other restrictions can be dropped, because (A) computes only feasible values of Q and V.

- Objective Function:

$$\frac{g \cdot \rho}{3600000} \sum_{k=1}^{N} \int_{t_{k-1}}^{t_k} a(t)Q(t)F(V(t))\eta(V(t),Q(t))dt \to Max!$$

5.4.2 <u>Numerical Results.</u> We use this model to solve the problems

(i) Optimization of the Gosau System during Phase II
     (see Figure 8)

(ii) Optimization of the Gosau System during the whole year
     (see Figure 9) (obviously, the model has to be extended,
     in this case).

In practice the knowledge of the optimal solution only is not satisfying. There are a lot of additional questions:
How does the production value depend on changes of the sinking process during Phase II?
How much does the optimal level of Lake Gosau from January until March depend on the influx Z?
In particular, we ask how does the optimal level of Lake Gosau on April 1st depend on the influx Z from April 1st until June 15th. We get the answer by solving a great number of the following problems:

Maximize the profit of the whole Gosau System during Phase II under the additional conditions:
The value of V_1 (April 1st) has to be within a given interval.
We solve this problem using the model above and adding the

Figure 8   Optimal level of Lake Gosau - Phase II

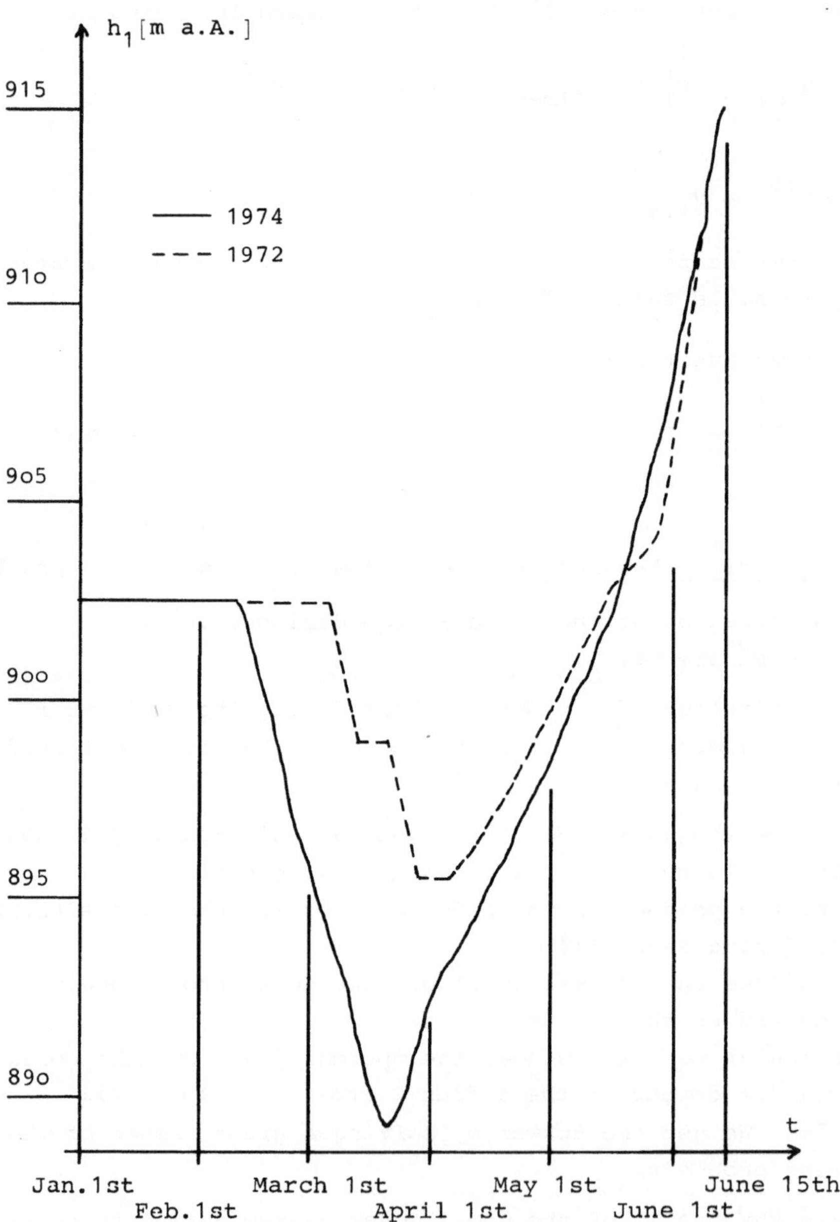

constraints

$$c_1 \leq V_1 (\text{April 1st}) \leq c_0$$

The solutions allow to present the optimal strategies of lowering Lake Gosau in dependence on the influx Z. This is very useful in practice: Knowing an approximation of $Z(t)$, $t \in$ [April 1st, June 15th], the manager can fix the level V_1(April 1st) and, therefore, the operation of the whole system during the time before April 1st. Moreover, the results show, that the estimation of Z need not be very good. For details see [2], [3].

<u>Figure 9</u>   Result of the annual optimization – level of Lake Gosau

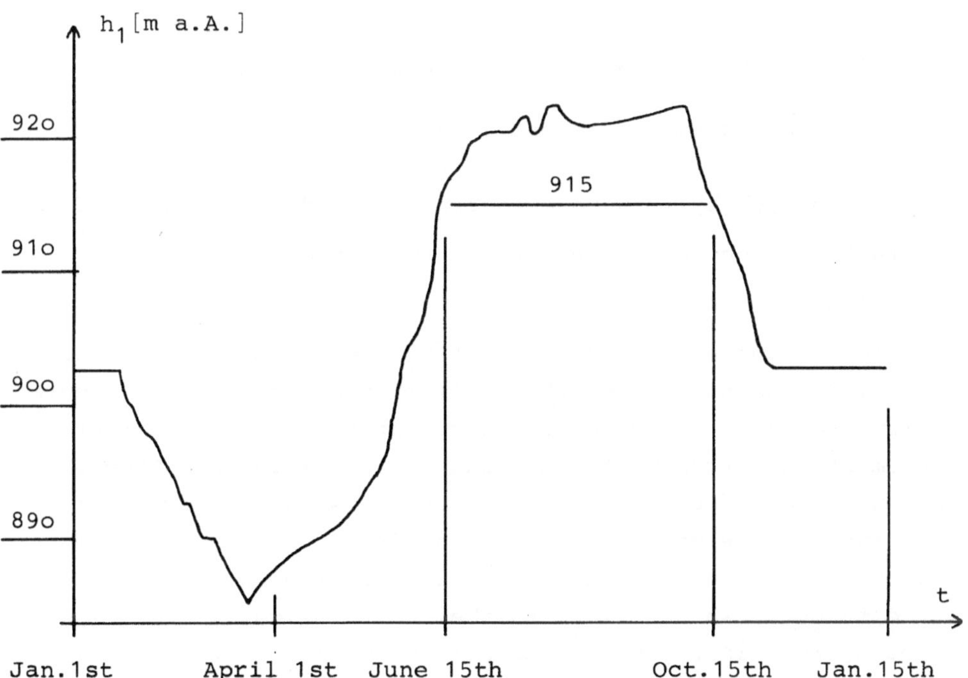

6   Determination of the Input Parameters

In (M) it is assumed that all input parameters are known. A certain amount of work was necessary to determine the seepage losses, the influces and the efficiency functions. Our considerations were based on careful measurements.

6.1   Seepage Losses Q_L

The seepage losses are of great influence on the hydrological situation of Lake Gosau and hence on the optimal operation mode. For the determination of Q_L we use the continuity equation

$$Q_L(V_1(t)) = Z_1(t) - Q_1(t) - \frac{dV_1}{dt}(t)$$

The available data were h_{1_k} and E_{1_k} where

h_{1_k} = height of Lake Gosau at day k in the morning

E_{1_k} = energy production of plant Gosau at day k

Approximating $\frac{dV_1}{dt}$ by forward differences and determining Q_1 from the data E_{1_k} we get an average value for the difference $Q_L(V_1(t)) - Z_1(t)$.

To eliminate the influence of Z_1 one needs measurements of Z_1 or, at least, methods which allow to estimate Z_1 from data, like temperature, precipitation, etc. But there exist neither measurements nor trustworthy methods of estimating Z_1. However, we are able to reduce the influence of Z_1 using the fact that the influx remains constant during the frost period (approximately November until March). Hence we determine $Q_L(V_1) - Z_1$ for those data only, which belong to the frost period.
Proceeding in this way we may expect to obtain a good approximation of Q_L. The result is shown in Figure 1o. For more details see Wacker et al [12].

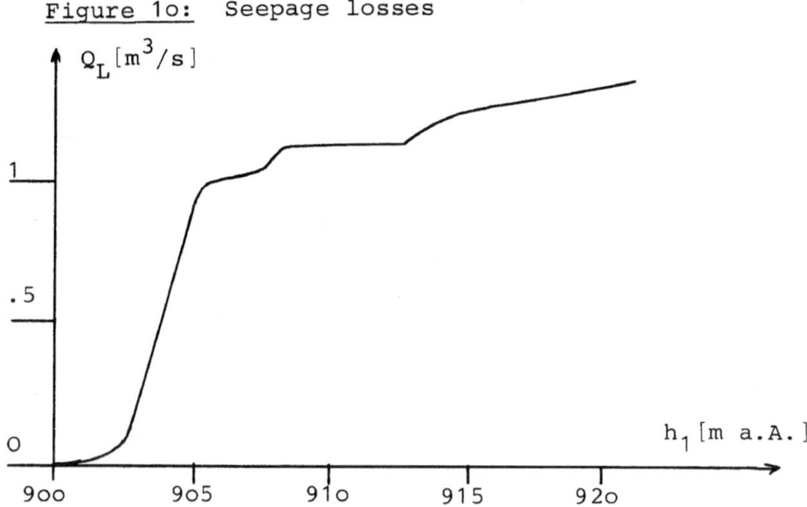

Remark: We present the seepage losses in dependence on the
height, because of historic reasons. To get this function we
evaluated the data of some 45 years (193o - 1974). The bends
could be fixed by the observation that some brooks start or stop
flowing when the level of Lake Gosau passes certain heights.

6.2   Influx Determination

We refer to 6.1, where the difficulty of the influx
prediction, resp. measurement in a karst region is mentioned.
To get realistic data for our optimization, we determine Z from
the production and the reservoir level data of the last years
(1972 - 1982), using the continuity equations. Now, knowing the
function Q_L, Z_1, Z_2 and Z_3 are simply determined by evaluating
the available data. For details see [3], [12].

Note: The determination of Z_2 and Z_3 may be inaccurate in some
cases, because we did not know the values of S_2 and S_3. However,
the number of these evaluations, which might be wrong, is
negligible. The management always tried to operate the system
such that $S_2 = S_3 = 0$.
As an example we present the influx data Z_1 of the years 1972
and 1974 (Figure 11).

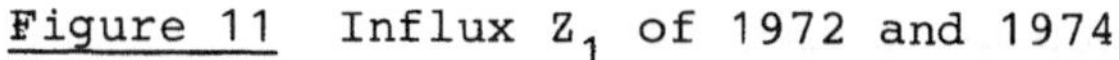

Figure 11   Influx Z_1 of 1972 and 1974

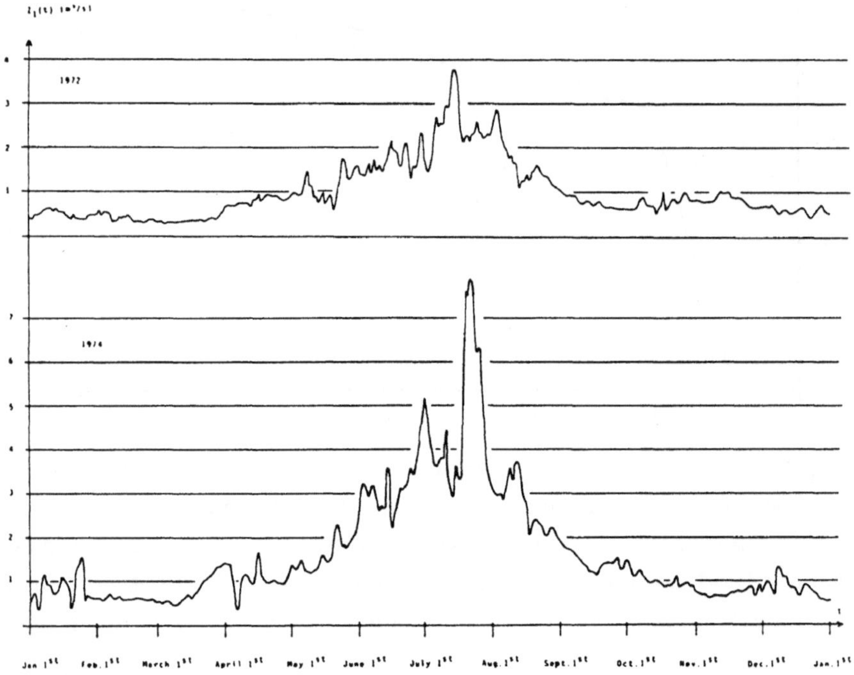

6.3   Efficiency Function

The efficiency functions η_i, $i = 1,2,3$ are required for evaluating the objective. As an example we present the determination of η_1 in this section. We drop subscript 1 for all parameters used in the following.

The efficiency function $\eta(f(V),Q)$ includes losses in the pipe system (i), in the turbines (ii), in the generators and the transformers (iii) and in the 3o kV (= 3oooo V) internal network (iv). For (iii) trustworthy data were presented by the producer, for (iv) we use actual measurements and Ohm's law. To determine the losses (i) and (ii) we use two different approaches

(a) Explicit approach. The fluid mechanics tells,that the losses in (i) depend quadratically on Q. Hence we use a quadratic polynomial in Q to approximate these losses. The coefficients are determined by the method of linear least squares,

using actual measurements. For determining the efficiency
function of the turbine we had two kinds of information; a curve
presented by the producer and actual measurements taken during
operating the power plant. Unfortunately, there was a significant discrepancy between them. Therefore, we proceeded as follows.

At first, the data presented by the producer were
approximated by the polynomial

$$\sum_{i=0}^{2} \sum_{j=0}^{3} a_{ij} f(V)^i Q^j$$

The coefficients a_{ij} are determined by the linear least squares
method.

Secondly we postulated that the sketch of the curve above
is correct but for an additional constant. Hence we minimized

$$\sum_{k=1}^{M} w_k (\eta(f(V_k),Q_k) - \eta_k)^2 + \sum_{i=0}^{2} \sum_{j=0}^{3} w_{ij} \left(\frac{b_{ij} - a_{ij}}{a_{ij}}\right)^2$$

with respect to b_{ij}, where

$$\eta(f(V),Q) = \sum_{i=0}^{2} \sum_{j=0}^{3} b_{ij} f(V)^i Q^j$$

$f(V_k), Q_k$ measurements $k = 1,\ldots,42 \ (= M)$

η_k efficiency factor of the turbine with respect
to measurement k

$w_k, \ w_{ij}$ weights with $w_{oo} := 0$

Result: The function η has the same shape as the curve of the
producer and differs 1 percent, at most, from the measurements.

(b) <u>Implicit approach.</u> The classical method of least
squares is used to solve problems with errors in the measurements.
In our case we have to take into account errors in the points
where the measurements are taken. Schwetlick/Tiller [11] proposed
a method, which also copes with this problem. We used this
method to determine a second approximation of the losses in (i)
and (ii). Now we had to solve a nonlinear system of some hundred

unknowns. Using the special structure of our problem, this can be done quite efficiently. A detailed description of this approach is given in [1o], this volume. The results have just the same accuracy as in (a).

7   Realization in Practice

        At the Gosau system no computer facilities have been available. Therefore, our results had to be formulated as operating rules.

7.1   Phase I

        We confine ourselves to give only the main points here. As far as possible we sought for rules which were independent of the actual influx situation.

  i) Start draining Lake Gosau exactly on October 1st when the higher winter tariff is valid.

 ii) The level of 915 m a.A. must not be reached before October 15th because of tourism. Depending on the starting level and the influx situation the intensity of the sinking process varies.

iii) After October 15th further draining is done rather quickly within about a month. There is even (a reduced) production during the night tariff period to increase the intensity of sinking and, therefore, to decrease the seepage losses.

 iv) It is essential not to stop sinking before the level of 9o2.5 m a.A. is reached. For lower levels the seepage losses are almost constant.

  v) For the other two plants there holds: avoid spillage water at all, if possible. Start production at Gosauschmied at 5 a.m to enable production at Steeg at 6 a.m.

 vi) For all plants: try to operate them near their optimal efficiencies.

        To demonstrate the efficiency of the mathematical optimization we present some data for Phase I. First hints were already used in 1982, final results have been available since 1983.

The influx situation is roughly described by the production
coefficient pc. pc less, and greater, than 1 indicate a dry, and
a wet, period, resp. So we also give the normalized values for
production (pc = 1).

year	production GWh	pc	$\dfrac{\text{production}}{\text{pc}}$	sinking time	start
1968	1.75	–	–	75	5.1o.
1969	1.55	o.39	3.97	12o	2o.o9.
1971	o.65	o.56	1.16	164	1o.o8.
1974	1.28	1.58	o.81	1o9	13.11.
1979	2.o9	1.11	1.88	65	15.1o.
1982	1.93	o.60	3.22	78	22.o9.
1983	2.65	o.78	3.4o	51	1.1o.
1984	2.55	o.60	4.25	38	2o.1o.
1985	2.26	o.46	4.91	46	5.1o.
1986	2.o4	o.75	2.72	66	29.o9.

Table 5:   Results in practice for Phase I

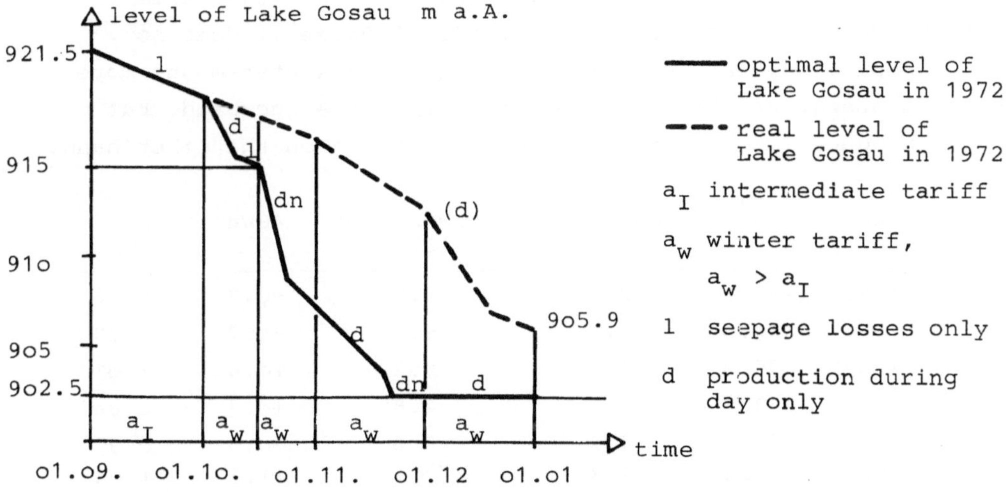

Figure 12: Results for Phase I in 1972

dn production during
day and night

(d) production during
part of day only

Comparing with past years the greatest difference consists in the operating rules i), iii), iv), cf. Tab. 5 and Fig. 12. There were years, for instance, when Gosau was almost inactive until November for purposes of reserve. Because of the seepage losses, however, most of the water was lost without producing energy.

7.2   Phase II

7.2.1 Simulation. By simulation it was possible to answer the following question. Starting at January 1st, is it preferable to operate the plant at the level of 9o2.5 m and then, say in February, to sink the reservoir or vice versa? The answer is slightly in favour of the first strategy. Further simulation explains that pumping should start as late as possible even if it must be done at the price of the higher day tariff.

7.2.2 Optimization. Here one has to estimate the lowest level h^* (April 1st) of Lake Gosau. This can be done approximately by the experienced managing engineer taking into account rainfall, snow etc. Production reacts very sensitivly on wrong estimations but the value of the production of all of Phase II does not. For example, a lower level for Lake Gosau in winter means more pumping energy at the low summer tariff on the one hand, but Steeg produces more at the high winter tariff on the other hand.

year	h^*	production [GWh]	consumption [GWh]	ΔGWh	pc
1969	892.5	2.5	2.7	−o.2	o.78
1976	898.o	1.o	1.7	−o.7	o.79
1983	896.o	1.6	1.2	+o.4	1.o1
1984	887.9	3.5	2.8	+o.7	o.86
1985	886.3	4.2	3.6	+o.6	o.9o
1986	885.8	4.3	3.1	+1.2	1.o8
1987	884.7	5.1	3.5	+1.6	1.31

Table 6:   Results in practice for Phase II
             ΔGWh = consumption − production

7.3 Pumping Activities

In case of surplus energy, mostly produced by river power plants, cheap energy is available. This energy may be used to pump water back from the reservoir Gosauschmied to Lake Gosau. It was possible to answer the questions:

i) The managing eingineer gets an offer for pumping energy. Should he accept or not?

ii) Spillage water at Klaushof can be avoided by pumping water from the reservoir Gosauschmied to Lake Gosau. At what price for pumping energy does it pay?

We worked out a table where the marginal costs were listed in dependence on the level of Gosau and the selling price of energy. This table was concentrated into a rule of thumb to simplify application. So far no figures are available.

Acknowledgement:

The data were made available to us by the OKA. Thanks are given to the management of the OKA, Dir.Dipl.-Ing.W.Barwig, Ing.R.Peßl, and Ing.E.Feichtinger for their substantial help which includes laborious experiments. Our practical work was supported both by the Austrian Ministry of Science and Research and by the OKA. We thank Min.Rat.Dr.O.Zellhofer for his help to manage the contract with the OKA. Our theoretical work has been supported partly by the Austrian Science Foundations Fonds under grants P4786, and P5729.

References:

[1] W.Barwig and R.Peßl: Optimization of the Gosau system, in:
     Applied optimization techniques in energy problems,
     Hj.Wacker (ed.), B.G.Teubner, Stuttgart, 1985, 94-118

[2] W.Bauer: Modelling and optimization of the Gosau hydro
     electric power system, ZOR 3o(1986),B169-B2o4

[3] W.Bauer: Optimal control of storage power plants, VWGÖ,
     Wien 1987

[4] H.Gfrerer: Global konvergente Optimierungsverfahren mit
     Anwendung auf die Höhensteuerung eines Speicherkraftwerkes,
     VWGÖ, Wien 1983

[5] H.Gfrerer: Optimization of hydro energy storage plant
     problems by variational methods, ZOR 28(1984), B87-B1o1

[6] H.Gfrerer: Globally convergent decomposition methods for
     nonconvex optimization problems, Computing 32(1984),
     199-227

[7] H.Gfrerer: Optimization of storage plant systems by
     decomposition, in: Applied optimization techniques in energy
     problems, Hj.Wacker (ed.), B.G.Teubner, Stuttgart, 1985,
     215-226

 [8] H.Gfrerer, J.Guddat and Hj.Wacker: A globally convergent
      algorithm based on imbedding and parametric optimization,
      Computing 3o(1983), 225-252

 [9] H.Gfrerer, J.Guddat, Hj.Wacker and W.Zulehner: Path-following
      methods for Kuhn-Tucker curves by an active index set
      strategy, in: Systems and Optimization, A.Bagchi and H.Th.
      Jongen (eds.), Lecture Notes in Control and Information
      Sciences 66, Springer, Berlin, 1985, 111-131

[1o] E.H.Lindner and Hj.Wacker: A black box technique for
      determining the efficiency function of a hydroelectric
      storage power plant, this volume

[11] H.Schwetlick and V.Tiller: Numerical methods for estimating
      parameters in nonlinear models with errors in the variables,
      Technometrics 27(1985)1, 17-24

[12] Hj.Wacker, W.Bauer, H.Gfrerer, E.Lindner, A.Schwarz:
      Erzeugungsoptimierung in Wasserkraftwerken (Gosaukraftwerks-
      kette), Arbeitsbericht für das BMfWuF und die OKA, 1985,
      1-22o

A BLACK BOX TECHNIQUE FOR DETERMINING THE EFFICIENCY FUNCTION OF A HYDROELECTRIC STORAGE POWER PLANT

Ewald H. Lindner and Hansjörg Wacker

1 Introduction

In the frame of hydro energy optimization an essential problem is to determine certain input parameters, e.g. the total efficiency functions of the power plant, the influx to the reservoir etc. In this paper we confine ourselves to describing a new technique for determining the efficiency function. The second problem is treated in [1o], [11]. In section 2 we give a basic mathematical model for hydro energy optimization, including the mathematical formulation of the hydrological background. In section 3 two failures for determining the efficiency function are sketched. In section 4 we present a special technique of least squares type. Actually, we have a set of measurements, where not only the measurements but also the measure points are subject to errors. Based on these measurements the efficiency function is determined. The standard nonlinear least squares technique does not cope with this problem. Hence, we use a generalized version proposed by Schwetlick/Tiller [12] in combination with regularization. Our technique was successfully applied to the storage plants Partenstein and Gosau, both situated in Upper Austria. This technique easily carries over to any other problem of this type, i.e., determining some "transfer function" by means of measurements, where the measure points are also subject to errors.

2   A Mathematical Model for a Hydroelectric Storage Power Plant

Figure 1 gives the essentials of a storage power plant system. The influx Z runs into the reservoir, whose content is V. We consider a discharge Q which passes through the conducting system (head rise gallery, surge tank, penstock) down to the turbines. The potential energy of Q is changed into kinetic energy (water → turbine) and afterwards into electric energy (turbine → alternator), which is then transformed (transformer).

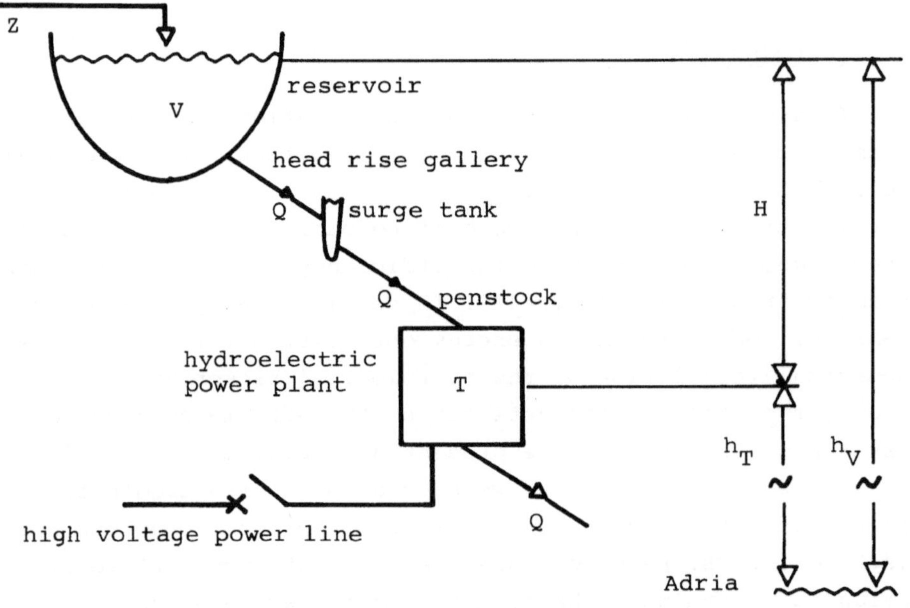

Figure 1: Diagram of a hydroelectric power plant with one turbine

For a power plant with one turbine this can be formalized as follows:

$$(1) \quad P_{Th} = g \cdot \rho \cdot H \cdot Q \qquad \text{theoretic power} \quad [Ws]$$

$$(2) \quad P = P_{Th} \cdot \eta(H,Q) \qquad \text{actual power} \quad [Ws]$$

i.e., the theoretic power P_{Th} is reduced by the efficiency $\eta \in (0,1)$ due to losses during transforming the energy. (Notations see below.)

For a power plant with n turbines with discharges $Q_1, \ldots, Q_n$ we have

$$(3) \qquad P_{Th} = g \cdot \rho \cdot H \cdot \sum_{i=1}^{n} Q_i = \sum_{i=1}^{n} g \cdot \rho \cdot H \cdot Q_i = : \sum_{i=1}^{n} P_{Th,i}$$

$$(4) \qquad P := \sum_{i=1}^{n} P_i := \sum_{i=1}^{n} P_{Th,i} \cdot \eta_i (H, Q_1, \ldots, Q_n)$$

Our main aim is to maximize the monetary income of the power plant within a fixed interval of time $[T_B, T_E]$. Hence, we use the following model (M):

Model (M):

Maximize

$$(5) \qquad E(T_B, T_E) = \int_{T_B}^{T_E} \frac{a(t)}{3\ 600\ 000} \sum_{i=1}^{n} g \cdot \rho \cdot H(t) Q_i(t)$$

$$\eta_i (H(t), Q_1(t), \ldots, Q_n(t)) dt$$

subject to

$$(6) \qquad V'(t) = Z(t) - \sum_{i=1}^{n} Q_i(t) - S(t) \qquad \text{a.e. in } [T_B, T_E]$$

reservoir continuity equation

$$(7) \qquad V(T_B) = V_B, \qquad V(T_E) = V_E$$

$$(8) \qquad V_{min} \leq V(t) \leq V_{min}$$

$$(9) \qquad Q_{min,i} \leq Q_i(t) \leq Q_{max,i} \qquad i = 1, \ldots, n$$

$$(10) \qquad 0 \leq S(t) \leq S_{max}$$

$$(11) \qquad V(t) = C(h_V(t)), \quad h_V(t) = H(t) + h_T$$

for $t \in [T_B, T_E]$

We use the following notations (for $t \in [T_B, T_E]$ and $i = 1, \ldots, n$)

$E(T_B, T_E)$	total rated energy production in the interval $[T_B, T_E]$, ATS ... Austrian Schilling	[ATS]
$[T_B, T_E]$	time interval under consideration	[s]
$a(t)$	tariff at time t, cf. [3] this volume	[ATS/kWh]
$P_i(t)$	actual power of power set i at time t	[Ws]
$P_{Th,i}(t)$	theoretic power of power set i at time t	[Ws]
$\eta_i(H, Q_1, \ldots, Q_n)$	total efficiency of power set i	[-]
g	acceleration due to gravity, 9.81 m/s^2	
ρ	density of water, 1ooo kg/m^3	
$H(t)$	head at time t	[m]
$h_V(t)$	level of the reservoir a.A. (above Adria)	[m]
h_T	level of the turbines a.A.	[m]
$Q_i(t)$	discharge through turbine i at time t	$[m^3/s]$
$V(t)$	volume of water in the reservoir at time t	$[m^3]$
$z(t)$	influx into the reservoir at time t	$[m^3/s]$
$S(t)$	spillage water at time t	$[m^3/s]$
V_B, V_E	initial, and terminal content of the reservoir, resp.	$[m^3]$
C	reservoir characteristic	$[m^3]$

Subscripts min, and max, indicate given lower, and upper, bounds, resp.

The function C is smooth and strictly increasing, hence, its inverse C^{-1} exists and is also smooth. The electric power plants company in Upper Austria, OKA, is interested to increase its income which just means to maximize $E(T_B, T_E)$. For more details, on (M), on $a(t)$, etc., cf. [3], this volume. In addition we also investigated different objectives, such as the problem to

- maximize the energy produced within a fixed interval of time
  under restrictions similar to those described above, cf.[4],[5]

- maximize the power output when the amount of water to be consumed is prescribed, cf. [1]

- minimize the consumption of water when the power output is prescribed, cf. [1] (peak power demands)

In any case, the efficiency function is an essential part for calculating either the objective or the constraints.

We want to apply our technique to the power plants Partenstein and Gosau, both situated in Upper Austria. Therefore, let us present the relevant data. Under some acceptable simplifications the hydroelectric power plant Partenstein may be modelled by (M). There are n = 2 Francis turbines with vertical shaft, 16.2 MW power and Q_{max} = 13 m^3/s maximum consumption each, and

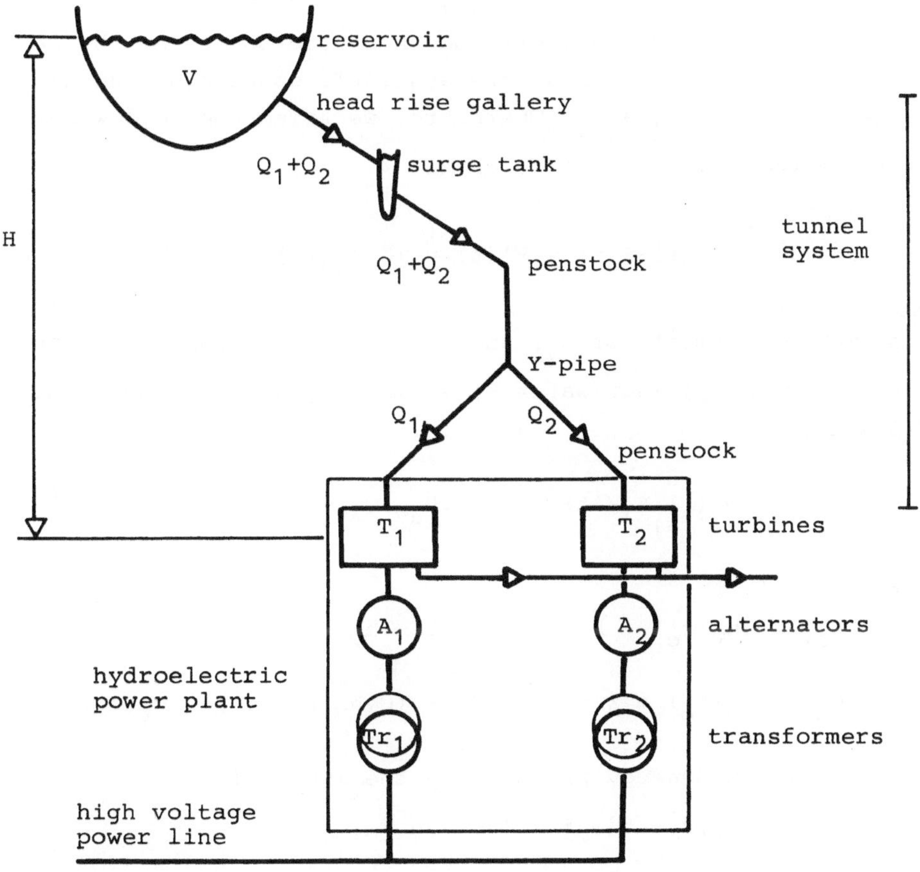

Figure 2: Sketch of the power plant Partenstein

2 three-phase synchronous alternators with 21.5 MVA, 55oo V,
5o Hz each. Each power set is completed by a 215oo kVA,
55oo/11oooo V transformer. The maximum content of the reservoir
is 736ooo m^3, its level varies from 451.oo to 456.2o m.
h_T is 291.oo m. As the level of the reservoir should not sink
below 452.2o m we confine our interest to heads between 161.2o m
and 165.2o m. For the data of the hydroelectric power plant Gosau
see [3], this volume.

3   Determining the Efficiency Function: Two Failures

3.1   A First Attempt (Partenstein)

The total efficiency η_{total} of the power plant is defined
as the ratio of produced power and theoretic power. At the beginning of our work no data were available concerning the discharges Q_1, Q_2 or $Q_1 + Q_2$. Therefore, we tried the following
ansatz. From (6), and (11) we obtain

$$Z(t) - \sum_{i=1}^{n} Q_i(t) - S(t) = V'(t) = (C(h_V(t)))'.$$

With two measurements of h_V, and Z, at time t_o, and $t_o + \Delta t$, and
if there is no spillage water S, we may approximate the total
efficiency by

$$\eta_{total} \sim (\sum_{i=1}^{n} E_i / \Delta t) \ / \ (g \cdot \rho \cdot H_{av} \cdot Q_{av})$$

where

$$H_{av} = (h_V(t_o) + h_V(t_o + \Delta t))/2 - h_T$$

$$Q_{av} = (Z(t_o) + Z(t_o + \Delta t))/2 - (C(h_V(t_o + \Delta t)) - C(h_V(t_o)))/\Delta t$$

E_i is the energy produced in the interval

$$[t_o, t_o + \Delta t], \ i = 1, \ldots, n$$

If there is no information on the discharges there seems to be
no other chance for modelling. For $n \geq 2$ there is at least one
major drawback: this ansatz does not contain the distribution of
the total discharge to the n turbines. This problem can be overcome only partly by taking measurements when just one turbine is
in operation. However, even for determining the efficiency of a
single turbine our numerical results do not recommend this
method. This is due to the fact Q_{av} reacts very sensibly to
small errors in the measurements of h_V, as $Z \sim 1o$, but
$C'(h_V) \sim 1.4*1o^5$. This behaviour for Partenstein is quite typical: Kühne [9] analysized 38 Swiss reservoirs whose $C'(h_V)$
values ranged from $5*1o^5$ to $13*1o^6$. Some Austrian reservoirs
we are informed on have values between $1o^5$ and $1o^6$.

3.2   A Second Attempt (Partenstein and Gosau)

This time we used a model motivated by engineering considerations. The theoretic power is reduced by losses

  i) in the tunnel system conducting the water from the reservoir
    to the turbines

 ii) in the turbines themselves, and

iii) in the alternators and transformers.

With respect to the power plant Partenstein the produced electric
energy is no more under its responsibility afterwards. Therefore,
we do not consider further losses in the network etc. However,
for the power plant Gosau we include losses

 iv) in the 3o kV power line

from the power plant itself to the 3o kV/11o kV power line interface.

Let us now consider i) - iv) in detail. For the hydrological situation cf. Fig. 2.

  i) The water is brought down to the power plant by a single
    tunnel (head rise gallery, penstock) which bifurcates before
    the turbines. By engineering assumptions the losses of

pressure in the tunnel due to friction, turbulences, etc. may be expressed by a fictitious reduction of the head H which is proportional to the square of the discharges in each section of the tunnel system. We get

(12) $\quad H^{(2)} := H - c(Q_1 + Q_2)^2 \qquad$ in the head rise gallery and in the penstock before the Y-pipe

(13) $\quad H_i^{(3)} := H^{(2)} - c_i^{(1)} [((Q_1 + Q_2)/F_o)^2 - (Q_i/F_i)^2]$

$$\text{in the Y-pipe for } i = 1,2$$

(14) $\quad H_i := H_i^{(4)} := H_i^{(3)} - c_i^{(2)} Q_i^2 \qquad$ in the penstock between the Y-pipe and the turbine, for $i = 1,2$

Here F_o denotes the circular cross-section of the penstock before the Y-pipe. F_1, and F_2, denote the cross-section of the penstock after the Y-pipe towards turbine 1, and 2, resp., cf. Fig. 3.

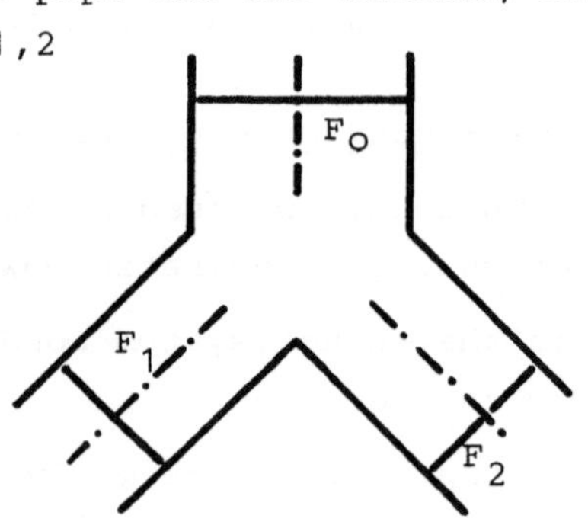

Figure 3: Y-pipe

Because of the given special construction we may assume $c_1^{(1)} = c_2^{(1)} =: c^{(1)}$. Combining (12) – (14) results in

(15) $\quad H_i = H - c_o(Q_1 + Q_2)^2 - c_i Q_i^2$

for the manometric head H_i before turbine i, for $i = 1,2$. There is $c_o := c + c^{(1)}/F_o^2$ and $c_i := c_i^{(2)} - c^{(1)}/F_i^2$ for $i = 1,2$. Therefore, the actual power before turbine i is

(16) $\quad P_{bT,i}(H, Q_1, Q_2) := g \cdot \rho \cdot H_i Q_i = g \cdot \rho \cdot H Q_i \eta_{s,i}(H, Q_1, Q_2)$

where

(17) $\quad \eta_{S,i}(H,Q_1,Q_2): = 1-c_o(Q_1+Q_2)^2/H-c_iQ_i^2/H \quad$ for $i = 1,2$

By physical interpretation c, $c_i^{(j)}$, $i,j = 1,2$ are positive. Note, that c_1, c_2 are not necessarily positive any more. c, $c_i^{(j)}$, $i,j = 1,2$ are so-called friction coefficients, tabulated in engineering handbooks.

ii) As for Partenstein there are two identical turbines we assume that the efficiency functions $\eta_{T,1}$, and $\eta_{T,2}$ for turbine 1, and 2, are identical, denoted by η_T. The producer of turbines presents η_T as a function depending on H_i, and Q_i. Sketched in an H_i-Q_i-diagram the contours of η_T are essentially shell-like, cf. Fig. 4.

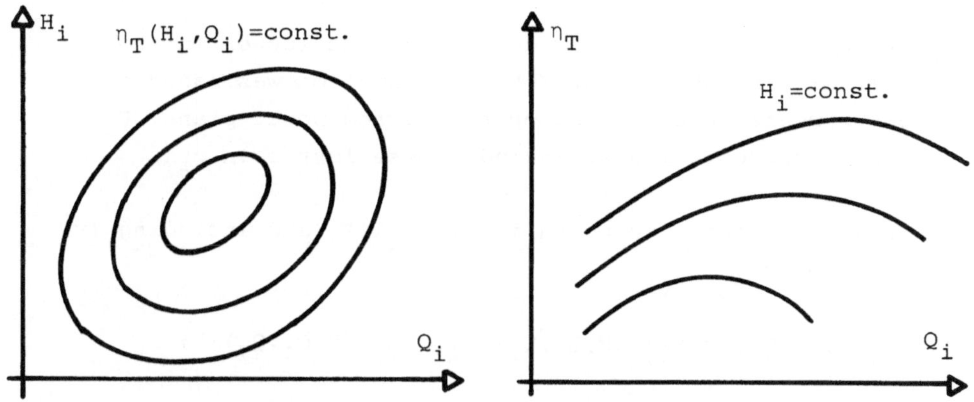

Figure 4: Efficiency function of a turbine

Therefore, we use a bi-quadratic approximation, i.e.

(18) $\quad \eta_T(H_i,Q_i) = a_o+a_1Q_i+a_2H_i+a_3H_iQ_i+a_4Q_i^2+a_5H_i^2$

where a_j, $j = o,\ldots,5$, are determined by linear least squares.

For the output power of turbine i we obtain

(19) $\quad P_{T,i}(H,Q_1,Q_2) := P_{bT,i}(H,Q_1,Q_2) \cdot \eta_T(H_i,Q_i) =$

$$= g \cdot \rho \cdot HQ_i \eta_{S,i}(H,Q_1,Q_2) \cdot \eta_T(H\eta_{S,i}(H,Q_1,Q_2),Q_i)$$

iii) Again for Partenstein both alternators and transformers are
   identical. The producers presented curves for the efficiencies in dependence on the relevant output powers.
   However, it is no problem to get the dependence on the input
   powers. A quadratic approximation, its coefficients deter-
   mind by linear least squares, turned out to be sufficient
   for both the alternator and the transformer. As calculating
   the efficiencies for the alternator and the transformer was
   done separately we needed 6 coefficients. To approximate
   the joint efficiency $\eta_{AT}(P_{T,i})$, $i = 1,2$ just 5 parameters
   were needed.

 iv) For Gosau we additionally determined the losses in the
   power line by Ohm's law. The approximation was based on
   measurements taken during operating the power plant. For
   ease of presentation we include these losses in η_{AT}.

   Following the line given in i) - iii) (and including iv)
for Gosau) we get:

(2o) $\quad P_i(H,Q_1,Q_2) = P_{T,i}(H,Q_1,Q_2) \cdot \eta_{AT}(P_{T,i}(H,Q_1,Q_2)) =$

$$= g \cdot \rho \cdot HQ_i \eta_{S,i}(H,Q_1,Q_2) \cdot \eta_T(H\eta_{S,i}(H,Q_1,Q_2),Q_i)$$

$$\cdot \eta_{AT}(g \cdot \rho \cdot HQ_i \eta_{S,i}(H,Q_1,Q_2) \cdot \eta_T(H\eta_{S,i}(H,Q_1,Q_2),Q_i))$$

resp.

(21) $\quad \eta_{Total}(H,Q_1,Q_2) = \dfrac{P_1(H,Q_1,Q_2)+P_2(H,Q_1,Q_2)}{P_{Th}(H,Q_1+Q_2)} =$

$$= \sum_{i=1}^{2} \frac{Q_i}{Q_1+Q_2} \eta_{S,i}(H,Q_1,Q_2) \eta_T(H_i,Q_i) \eta_{AT}(P_{T,i})$$

Hence, by combining the efficiency functions of all separate parts one easily obtains a total efficiency function. However, there are some major drawbacks.

- The friction coefficients are taken out of handbooks, their tabulated values range over quite a large region.

- The efficiency function of the turbines is calculated by the producer. Calculations are mainly based on measurements taken in a laboratory using a small scaled model. The shell-like curves are then updated by hydraulic equivalence formulas using measurements in the plant for one chosen standard head.

- η_{Total} calculated by this method differs by about 1o % from that efficiency the managing engineers estimated by experience.

- Even if the friction coefficients and the efficiency function of the turbine were correctly determined at a certain time, they are changing due to aging. Above all the head rise gallery and the turbine blades are not in the same condition as at that time they were constructed.

4   Determining the Efficiency Function: A Black Box Technique

In the following we discuss a black box model. This means, we consider the head H and the discharges as inputs and the produced power(s) P as output(s). We want to determine a transfer function T so that $P = T(H,Q_1,Q_2)$. To include the experience of the engineers we choose the class of functions for T to be of the form (21). This procedure ends up with an identification problem for all those parameters mentioned in i) - iv) of section 3.2. In principle solution could be performed by nonlinear least squares based on measurements for P^j at the points (H^j,Q_1^j,Q_2^j).

All measurements have to be taken during operation. To reduce the number of parameters we proceed as follows. Due to accurate measurements we can predetermine η_{AT}, and get rid of iii) (and iv)) of section 3.2. The power produced by the power set i can easily be measured between the alternators and the transformers, which will be denoted by $P_{A,i}$. Hence, our black box technique for determining the total efficiency function of a hydroelectric power plant consists of the following steps:

- Determine the efficiency functions for the alternators and transformers separately (and for the high voltage power line).

- Take l measurements of H, Q_1, Q_2, $P_{A,1}$, $P_{A,2}$ denoted by H^j, Q_1^j, Q_2^j, $P_{A,1}^j$, $P_{A,2}^j$, $j = 1,\ldots,l$, during operating the power plant. $P_{A,1}$, $P_{A,2}$ are measured after the alternator, but before the transformer (see Fig. 2).

- Determine the parameters $(c,a) := (c_o,c_1,c_2,a_o,\ldots,a_5)$ for $\eta_{S,i}$, and η_T, cf. (17), and (18), resp., based on these measurements.

- As in section 3.2, the total efficiency is now given by (21).

Classical least squares presume that the points (H,Q_1,Q_2) where the measurements were taken, are exact, which is not the case here. For Partenstein Q_1, Q_2 can only be measured subject to an error of up to 1o %. In the case of Gosau the error was substantially reduced by use of a Venturi tube. Concerning the head H the error is relatively small but the result is very sensible to any variation. However, this problem can satisfactorily be treated by the following technique proposed by Schwetlick/Tiller [12].

4.1 The nonlinear least squares method of Schwetlick/Tiller

Consider the nonlinear system

$$y = r(x,\theta)$$

with input $x \in \mathbb{R}^k$, output $y \in \mathbb{R}^m$, and parameters $\theta \in \mathbb{R}^p$ characterising the system. We want to identify the parameters θ by measurements (x^j,y^j), $j = 1,\ldots,l$ of both the independent variable x and the dependent variable y. Let r be continuously differentiable.

Including the errors of the independent variables (x^j-x^j) in the classical regression model, we obtain the following unconstrained optimization problem

Minimize

$$(22) \qquad \frac{1}{2} \sum_{j=1}^{1} \begin{pmatrix} r(x^j,\theta)-Y^j \\ x^j-X^j \end{pmatrix}^T W_j \begin{pmatrix} r(x^j,\theta)-Y^j \\ x^j-X^j \end{pmatrix}$$

with respect to $z: = (x^1,\ldots,x^1,\theta) \in \mathbb{R}^{kl+p}$.

The weights W_j are positive definite matrices.

Decomposing W_j into $R_j^T R_j$ by Cholesky factorization we can rewrite (22) as

$$(23) \qquad \text{Minimize } S(z): = \frac{1}{2}\|F(z)\|_2^2$$

where

$$F(z): = \begin{pmatrix} F_1(z) \\ F_2(z) \end{pmatrix} : = \begin{pmatrix} R_{11}(x^1-X^1) \\ \vdots \\ R_{11}(x^1-X^1) \\ ------- \\ R_{12}(x^1-X^1)+R_{13}(r(x^1,\theta)-Y^1) \\ \vdots \\ R_{12}(x^1-X^1)+R_{13}(r(x^1,\theta)-Y^1) \end{pmatrix}$$

and $R_j = \begin{pmatrix} R_{j3} & R_{j2} \\ O & R_{j1} \end{pmatrix}$ with R_{j1}, R_{j3} right upper triangular matrices.

Thus we arrive at a minimum norm problem. The Jacobian J_F is of dimension $(lk+p, lk+lm)$ - in the special case of Partenstein with 26 measurements this gives a $(26*3 + 9, 26*(3+2))$ matrix. As a first step we present the principal solution line which is just nonlinear least squares. In the second step we exploit the special structure of this type of problem.

a) Choose the initial point $z_o \in \mathbb{R}^{lk+p}$, i: = 0

b) Compute $J_F(z_i)$ and the gradient $\nabla S(z_i) = J_F^T(z_i) \cdot F(z_i)$

c) If $\| \nabla S(z_i) \|_2$ is sufficiently small, accept z_i as a minimizer
   of S, stop.

d) Solve the linearized problem (numerically):
   Minimize $\| J_F(z_i) \Delta z - F(z_i) \|_2^2$ with respect to Δz. Let the
   minimizer be Δz^*.

e) Perform a(n incomplete) line search for the Gauß-Newton-
   direction. Alternatively, any damping method may be used, cf.
   Deuflhard [6]. Let the minimizer be γ_i and $\Delta z_i : = \gamma_i \cdot \Delta z^*$.

f) If $\| \Delta z_i \|_2$ is too small, there will be no convergence, stop.
   Otherwise $z_{i+1} : = z_i - \Delta z_i$, i: = i+1, goto a)

The numerical effort for d) can be reduced substantially using
the following procedure. Here we profit by the special sparse
structure of the Jacobian J_F.

$$J_F(z) = J_F(x,\theta) = \begin{pmatrix} J_{11} & O \\ J_{21} & J_{22} \end{pmatrix} \in \mathbb{R}^{lk+p}_{l(k+m)}$$

where $J_{11} \in \mathbb{R}^{lk}_{lk}$ is block diagonal, each diagonal block
$J_{11,j} = R_{j1} \in \mathbb{R}^k_k$ is an upper triangular matrix, for j = 1,...,l.
$J_{21} \in \mathbb{R}^{lk}_{lm}$ is block diagonal; each diagonal block $J_{21,j} \in \mathbb{R}^k_m$ is
full, for j = 1,...,l, and $J_{22} \in \mathbb{R}^p_m$ is full.

By an orthogonal transformation Q (Householder) we decouple Δx,
and $\Delta \theta$ in d)

(24) $\| J_F \cdot \Delta z - f \|^2 = \| Q(J_F \Delta z - f) \|^2 = \| K_{11} \Delta x + K_{12} \Delta \theta - g_1 \|^2 +$

$$+ \| K_{22} \Delta \theta - g_2 \|^2$$

where Q is determined so that

$$QJ_F = K = \begin{pmatrix} K_{11} & K_{12} \\ O & K_{22} \end{pmatrix}, \quad g := \begin{pmatrix} g_1 \\ g_2 \end{pmatrix} := Qf$$

where K_{11} is block diagonal. Its diagonal elements $K_{11,j} \in \mathbb{R}^k_k$ are upper triangular matrices, for $j = 1,\ldots,l$. Q is calculated via solving l separate linear least squares problems, each of dimension k, by Householder's method. As W_j is positive definite, $K_{11,j}$ is regular. Therefore, there exists a unique solution Δx for the first term in the sum (24) dependent on $\Delta\theta$. In detail,

$$K_{11}\Delta x = g_1 - K_{12}\Delta\theta, \quad \text{resp.}$$

$$K_{11,j}\Delta x_j = g_{1,j} - K_{12,j}\Delta\theta \quad \text{for } j = 1,\ldots,l,$$

i.e. one has to solve l linear triangular systems of dimension k each. Of course, one first has to determine $\Delta\theta^*$. This is performed by solving the p-dimensional problem

$$\text{Minimize } \|K_{22}\Delta\theta - g_2\|^2 \text{ with respect to } \Delta\theta.$$

If K_{22} is not of full rank, we use the following regularization for (24):

$$\left\| \begin{pmatrix} \sqrt{1+\lambda^2}\, K_{11} & \sqrt{1+\lambda^2}\, K_{12} \\ O & K_{22} \\ O & \lambda I \end{pmatrix} \cdot \Delta z - \begin{pmatrix} g_1/\sqrt{1+\lambda^2} \\ g_2 \\ O \end{pmatrix} \right\|$$

For more details see [7], [8], [12].
We applied this method successfully both to Partenstein and to Gosau.

4.2   Application to Partenstein

For Partenstein we had $l = 26$ measurements. As the efficiency of the alternator proved to be constant for all these values, i.e., $\eta_A = 0.98$, we actually used the following model in the above context

$$x: = (H,Q_1,Q_2)$$

$$y: = (P_{A,1}/(g*\rho*0.98), \ P_{A,2}/(g*\rho*0.98))$$

$$\theta: = (c,a) = (c_0,c_1,c_2, \ a_0,\ldots,a_5)$$

$$r(x,\theta) = \begin{pmatrix} r_1(x,\theta) \\ r_2(x,\theta) \end{pmatrix}$$

where $r_i(x,\theta) = HQ_i\eta_{S,i}(H,Q_1,Q_2)\eta_T(H_i,Q_i)$.

For $\eta_{S,i}$, and η_T see (17), and (19).

Hence $k = 3$, $m = 2$, $p = 9$.

The measurements $(X^j,Y^j) = ((X_1^j,X_2^j,X_3^j), \ (Y_1^j,Y_2^j))$ are

$((H^j,Q_1^j,Q_2^j), \ (P_{A,1}^j/(g*\rho*0.98), \ P_{A,2}^j/(g*\rho*0.98))$ for $j = 1,\ldots,1$.

The weight matrices W_j are chosen as $\mathrm{diag}(1/(0.1*Y_1^j)^2,$

$1/(0.1*Y_2^j)^2, \ 1/(0.002*H^j)^2, \ 1/(0.08*Q_1^j)^2, \ 1/(0.08*Q_2^j)^2)$, i.e. as

the weighted reciprocal values of the variances.

We confine ourselves to presenting the measurements, cf. Fig. 5, the resulting parameters and the maximum deviations for adjusting the measurements, cf. [8].

Result: (after 2o iterations)

$$c_0^* = 0.1060*10^{-1}$$

$$c_1^* = -0.1283$$

$$c_2^* = -0.1249$$

$$a_0^* = 0.9907*10^2$$

$$a_1^* = 0.3098*10$$

$$a_2^* = -0.1329*10$$

$$a_3^* = -0.2026*10^{-1}$$

$$a_4^* = 0.1720*10^{-1}$$

$$a_5^* = 0.4478*10^{-2}$$

$$\|S(c^*,a^*)\| = 1.913$$

$$\|\nabla S(c^*,a^*)\| = 0.8814*10^{-6}$$

H^j	Q_1^j	Q_2^j	$\dfrac{P_{A,1}^j}{g*\rho*0.98}$	$\dfrac{P_{A,2}^j}{g*\rho*0.98}$
164.6775	1o.2185	1o.7ooo	1456.24o	1477.o43
164.3775	8.8275	9.523o	1352.223	1383.428
162.735o	11.235o	11.342o	1456.22o	1466.6oo
163.2325	12.o375	12.o91o	156o.257	156o.257
161.9725	8.2925	8.881o	1196.197	1248.2o5
164.255o	1o.7ooo	11.128o	1456.24o	1477.158
163.53oo	11.449o	12.626o	156o.257	157o.659
164.965o	7.95oo	9.1ooo	1183.46o	1379.819
165.12oo	3.75oo	6.4ooo	636.84o	1136.368
164.485o	1o.oo45	9.844o	1383.43o	1352.22o
163.51oo	8.o25o	9.3o9o	1o55.77o	1352.22o
162.82oo	9.63oo	9.79o5	1352.22o	1383.43o
162.83oo	8.72o5	8.3246	1248.21o	1196.2oo
163.5ooo	13.oooo	13.oooo	156o.257	156o.257
163.14oo	11.556o	11.5o25	1466.64o	1456.24o
164.oooo	1o.914o	1o.593o	1477.o4o	1456.24o
163.o1oo	11.35oo	7.35oo	156o.257	988.163
163.86oo	8.55oo	1o.7ooo	1144.189	1414.633
164.255o	1o.15oo	11.1ooo	1352.223	1456.24o
163.19oo	7.oooo	12.oooo	956.958	156o.257
163.o2oo	5.oooo	4.oooo	676.111	665.7oo
162.99oo	7.oooo	3.oooo	1o19.368	416.o7o
162.79oo	3.oooo	7.oooo	551.29o	977.76o
162.62oo	13.oooo	5.oooo	1674.676	655.3o8
162.56oo	13.oooo	7.oooo	1664.274	894.547
162.51oo	4.3ooo	4.8ooo	655.ooo	655.3oo

Figure 5: Measurements for Partenstein

Maximum differences between measurements and their adjusted values:

o.o7892	H
o.2247	Q_1
o.1665 with respect to	Q_2
68.52	$P_{A,1}/(g*\rho*0.98)$
53.o4	$P_{A,2}/(g*\rho*0.98)$

c_1^*, and c_2^*, differ by just 2.7 %. Though the construction is slightly different after the Y-pipe, this indicates, that $c_1 = c_2$ might be assumed in our model additionally.

4.3   Application to Gosau

There is just one power set, hence, the efficiency function simplifies to

$$\eta_S(H,Q) = 1-cQ^2/H$$

$$H_1: = H \cdot \eta_S(H,Q)$$

$$\eta_T(H_1,Q) = a_o+a_1Q+a_2H_1+a_3H_1Q+a_4Q^2+a_5H_1^2$$

$$P_A(H,Q) = g\rho HQ\eta_S(H,Q)\eta_T(H_1,Q)\eta_A(P_T)$$

For the analogous notations see section 3.2.
We proceed as in 4.2, with $k = 2$, $m = 1$, $p = 7$, $l = 43$ and result in

$$c^* = \text{o}.8356$$
$$a_o^* = \text{o}.3724*1\text{o}$$
$$a_1^* = -\text{o}.149\text{o}$$
$$a_2^* = -\text{o}.4499*1\text{o}^{-1}$$
$$a_3^* = \text{o}.2784*1\text{o}^{-2}$$
$$a_4^* = -\text{o}.14\text{o}2*1\text{o}^{-1}$$
$$a_5^* = \text{o}.1334*1\text{o}^{-3}$$
$$\|S(c^*,a^*)\| = \text{o.o}8356$$
$$\|\nabla S(c^*,a^*)\| = \text{o.o}1992$$

where we used the weight matrices

$$W_j: = \text{diag}(1/(\text{o.ooo}7*H^j)^2, \; 1/(\text{o}.5*Q^j)^2,$$
$$1/(\text{o.ooo}25*P_A^j/(g*\rho*\text{o}.98))^2)$$

For details see [7].

To check our results we compare our figures with those in [2]. There the efficiency functions η_S, and η_T, were determined separately, but based on additional measurements. η, and η_{BB}, give the efficiency by the method in [2], and by our black box method, resp., cf. Fig.6.

η	H[m]	$Q[m^3/s]$	η_{BB}
	155	6.1o	o.66
	15o	6.o3	o.667
$\eta = o.66$	144	5.87	o.66
	138	5.63	o.66
	132	5.19	o.67
	132	3.64	o.66
	155	5.68	o.71
	15o	5.4o	o.7o
$\eta = o.68$	146	5.51	o.68
	142	5.37	o.68
	136	4.97	o.68
	132.8	4.3o	o.68
	155	5.5	o.72
$\eta = o.71$	152	5.o	o.717
	146	4.8	o.7o
	141.4	4.23	o.68

Figure 6: Comparing η of [2] and η_{BB} of the black box method

The result of [2] proved to be slightly more accurate due to considering more measurements and a careful weighting technique. On the other hand our black box technique is simpler to perform once the software is available.

Acknowledgements:

Parts of this work was supported by the Austrian Science
Foundation Funds under grants P4786, and P5729. We thank the
management of OKA, Dir.Dipl.-Ing.W.Barwig, Ing.R.Peßl, and
Ing.E.Feichtinger, for their substantial help which includes
laborious experiments. Our practical work was supported both by
the Austrian Ministry of Science and Research (Min.Rat Dr.O.
Zellhofer) and OKA.

References:

[1]   W.Bauer, S.Buchinger, Hj.Wacker: Einsatz mathematischer
       Methoden in der Hydroenergiegewinnung, ZAMM 64 (1984),
       227-243

[2]   W.Bauer, H.Gfrerer, E.Lindner, A.Schwarz, Hj.Wacker:
       Optimization of the storage power plant system Gosau -
       Gosauschmied - Steeg, in: Applied optimization techniques
       in energy problems, Hj.Wacker (ed.), B.G.Teubner, Stuttgart,
       1985, 119-137

[3]   W.Bauer, E.H.Lindner, Hj.Wacker: Optimization of systems of
       hydro energy power plants, this volume

[4]   W.Bauer, H.Reisinger, Hj.Wacker: Höhensteuerung eines
       Tagesspeicherkraftwerkes (Schwarzach), ZOR 26(1982),
       B145-B167

[5]   S.Buchinger, H.Reisinger, Hj.Wacker: Tagesoptimierung eines
       Speicherkraftwerkes auf der Basis einer Momentanoptimierung,
       ZOR 25(1981), B139-B157

[6]   P.Deuflhard: Linear and nonlinear least squares computing
       in view of control applications, Preprint 87, SFB 123
       (Stoch.math.Modelle), Universität Heidelberg, September 1980

[7]   G.Dworschak, J.Mörwald: Gauß-Newton-Methode zur Bestimmung
       von Systemparametern in nichtlinearen Quadratmittelpro-
       blemen mit Fehlern in den Meßdaten, Seminarbericht,
       Institut für Mathematik, Universität Linz, 1984, 1-79

[8]   W.Grubauer: Optimierungsmodelle bei Hydroenergieproblemen,
       Diplomarbeit, Institut für Mathematik, Universität Linz,
       Dezember 1983

[9]   A.Kühne: Charakteristische Kenngrößen schweizerischer
       Speicherseen, Geographica Helvetica 33 (1978)4, 191-199

[1o] E.Lindner, G.Oberaigner, Hj.Wacker, A.Wakolbinger: The
      hydroelectric storage power plant Partenstein: Forecasting
      models for the influx and efficiency determination, in:
      Proceedings der 18. Jahrestagung "Mathematische Optimierung",
      Rostock/Diedrichshagen, GDR, 23.-27.3.1986, Seminarbericht
      Nr. 85, Sektion Mathematik, Humboldt-Universität zu Berlin,
      Oktober 1986, 59-69

[11] E.Lindner, Hj.Wacker: Input parameters for the optimization
      of the system Partenstein: Forecasting of the influx,
      determination of the efficiency function, in: Applied
      Optimization techniques in energy problems, Hj.Wacker (ed.),
      B.G.Teubner, Stuttgart, 1985, 286-32o

[12] H.Schwetlick/V.Tiller: Numerical methods for estimating
      parameters in nonlinear models with errors in the
      variables, Technometrics 27(1985)1, 17-24